Introduction to Aerospace Engineering

Introduction to Aerospace Engineering

Elizabeth Simpson

MURPHY & MOORE
www.murphy-moorepublishing.com

Published by Murphy & Moore Publishing
1 Rockefeller Plaza,
New York City, NY 10020, USA
www.murphy-moorepublishing.com

Introduction to Aerospace Engineering
Elizabeth Simpson

International Standard Book Number: 978-1-63987-327-2 (Hardback)

Cataloging-in-Publication Data

Introduction to aerospace engineering / Elizabeth Simpson.
p. cm.
Includes bibliographical references and index.
ISBN 978-1-63987-327-2
1. Aerospace engineering. 2. Aeronautics. 3. Engineering. I. Simpson, Elizabeth.
TL545 .I58 2022
629.132 52--dc23

Contents

Preface

Aerospace engineering is an interdisciplinary field of engineering that combines the principles of fluid mechanics, materials science, structural engineering, etc., to design and develop space crafts and aircrafts. The discipline further bifurcates into two major branches, namely, aeronautical engineering and astronautical engineering. While aeronautical engineering deals with creating aircrafts, astronautical engineering focuses on space technology to build rocket-ships and satellites. Avionics is another branch of aerospace engineering which is concerned with the electrical aspects of the subject. Modern disciplines like computational fluid dynamics which use the techniques of CAD and CAM along with finite numerical analysis are also being extensively used by upcoming engineers to reduce the cost and increase the efficiency of the designing process, making them an indispensible part of aerospace engineering. This book is a compilation of chapters that discuss the most vital concepts in the field of aerospace engineering. Most of the topics introduced in it cover new techniques and the applications of aerospace engineering. Those in search of information to further their knowledge will be greatly assisted by this book.

To facilitate a deeper understanding of the contents of this book a short introduction of every chapter is written below:

Chapter 1- The branch of engineering which focuses on the design, testing, development as well as the production of spacecraft, aircraft and the systems associated with them is known as aerospace engineering. The diverse aspects of aerospace engineering such as astrodynamics, propulsion, aerodynamics and control engineering have been thoroughly discussed in this chapter.

Chapter 2- The types of aircraft can be broadly classified into lighter than air systems, heavier than air systems, and gliders and sailplanes. Airships and balloons are a few types of lighter than air systems while helicopters and fixed wing aircrafts fall under heavier than air systems. This chapter discusses these types of aircraft in detail.

Chapter 3- The manufacturing process of aircrafts includes the design, fabrication, assembly, testing and inspection of different components of the aircraft. This chapter has been carefully written to provide an easy understanding of the varied facets of aircraft manufacturing process such as aircraft design, flight control system, aviation and aeronautics.

Chapter 4- The study of the motion of air especially when it is affected by a solid object is termed as aerodynamics. A few areas of study under it are lift force, aerodynamic drag, lift-induced drag, center of pressure, boundary layer and compressible flow. All these diverse concepts related to aerodynamics have been carefully analyzed in this chapter.

Chapter 5- There are numerous devices and methods which are used for the measurement of pressure in aerospace engineering. A few of them are McLeod gauge, Pirani gauge, thermocouple gauge and hot-filament ionization gauge. The topics elaborated in this chapter will help in gaining a better perspective about these devices used for pressure measurement.

I would like to share the credit of this book with my editorial team who worked tirelessly on this book. I owe the completion of this book to the never-ending support of my family, who supported me throughout the project.

Elizabeth Simpson

1

Aerospace Engineering: An Introduction

The branch of engineering which focuses on the design, testing, development as well as the production of spacecraft, aircraft and the systems associated with them is known as aerospace engineering. The diverse aspects of aerospace engineering such as astrodynamics, propulsion, aerodynamics and control engineering have been thoroughly discussed in this chapter.

Aerospace engineering is the branch of engineering behind the design, construction and science of aircraft and spacecrafts. It is divided into two major overlapping branches: aeronautical engineering and astronautical engineering. The first deals with crafts that stay within the Earth's atmosphere, while the second deals with crafts that operate outside of the Earth's atmosphere.

Aeronautical engineering was the original term before the broader term "aerospace" overtook it in usage, as flight technology advanced to include operating crafts in outer space. Aerospace engineering is largely about artefacts that fly - including airplanes, rockets, satellites or helicopters. It is also about how fluids flow, control systems, the strength of materials and improving strength. It also deals with how to design things so that they can be reliably built and maintained.

Aerospace engineers design, develop and test aircraft, spacecraft or missiles, and supervise their manufacture. Those who work with aircraft are called aeronautical engineers, and astronautical engineers are those working specifically with spacecrafts.

Aerospace engineers develop new technologies for use in aviation, defence systems and space exploration, and sometimes specialize in areas like structures, vehicle movement and control, propulsion systems, communications, and overall vehicle design. They also may specialize in a particular type of aerospace product, such as commercial aircraft, helicopters, military fighter jets, missiles, spacecrafts and rockets. These engineers might also be experts in celestial mechanics, aerodynamics, thermodynamics, acoustics, propulsion, or guidance and control systems.

Atmosphere and its Properties

The atmosphere can be generally defined as an invisible gaseous blanket around the earth held by earth's gravitational force. In one way or another, it affects all the entity exists on the earth - it is closely connected to our lives. Without this blanket, it would be unimaginably cold at night and unbearably hot during the day. Unlike other planets or extraterrestrial bodies, the planet earth is very unique in a sense that it is covered with atmosphere that provides optimum condition to support life to flourish. The earth's atmosphere are composed of two primary gases, viz., nitrogen and oxygen and water vapor that back up the life whereas carbon dioxide, hydrogen or helium gases are abundant with no water vapor in the atmospheres of other planets. Detailed study of the structure of earth's atmosphere is extremely important to gather knowledge regarding its origin, atmospheric composition and processes, different phenomena related to weather and climate, aviation, communication system etc.

If we look into the earth's atmosphere, it is seen that the atmosphere is not physically uniform throughout but has got considerable variations, especially in terms of pressure and temperature with height. An examination of vertical profile of the atmosphere reveals that it is comprised of a series of layers. These layers could be defined in several ways such as based on:

- The vertical temperature distribution.
- Electrical properties of the layer.
- The gaseous composition of it.

Layers of Atmosphere based on Vertical Temperature Profile

Air temperature throughout atmosphere shows complicated vertical profile and may be divided into five main distinguished layers, namely troposphere, stratosphere, mesosphere, thermosphere and exosphere. They are layered one atop the other and each one is very much needed to support and protect the life on the earth. The atmospheric also contains transition layers between two successive main layers. The nomenclature of these transitional layers zones contains the term – pause at its end. These are tropopause (from the end of troposphere and till the beginning of stratosphere), stratopause (boundary separating stratosphere from mesosphere), mesopause (transition region between mesosphere and thermosphere) and thermopause (between thermosphere and exosphere).

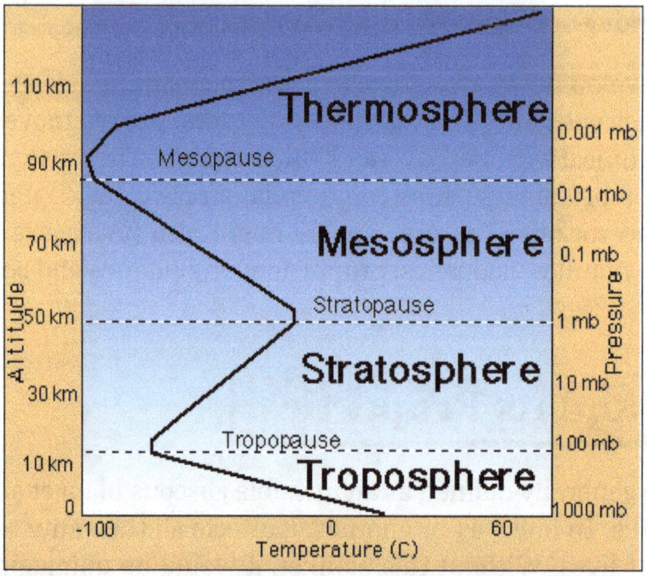

A schematic of the vertical profile of the atmosphere with relative position of its various layers. It also depicts the pattern of temperature changes with height from the ground and inside these layers. Approximate pressure levels at different altitude can also be noted.

Troposphere: Structure and Characteristics

The word 'troposphere' originates from the Greek word 'tropein' which means to change, circulate or mix. Troposphere is the lowest layer of the atmosphere which is close to the earth's surface where almost all the weather phenomenon that we experience like convection, turbulence and most of the cloud formation occur.

Thickness and Factors Affecting It

The thickness of this layer varies from 7-8 km at poles to 16-18 km over the equator. The depth of the troposphere is influenced by the latitude, season and time of the day. The reasons behind the non-uniformity in thickness of the troposphere are:

- Equator gets high insolation and strong convective fluxes: As equator gets highest amount of insolation, it becomes warmer than other parts of the earth. According to Charles' law, density reduces with increase in temperature at constant pressure which implies more the temperature over a place, more the air gets heated and rises upward due to decrease in density. Hence, due to differential heating of the earth's surface, equatorial regions experience more amounts of convective fluxes. Although both equator and poles have the same amount of atmospheric mass over it, the thermal expansion of air over equator is more than the poles and that makes the atmosphere thicker over equator than over the poles.
- Higher gravitational pull on atmospheric gases at poles: Due to ellipsoid shape of the earth, the gravitational pull is not uniform throughout and it is more over poles. Atmospheric molecules are drawn with more force near the poles causing contraction of the atmosphere.
- Equator has highest centrifugal force due to earth's rotation: The atmosphere over equator tends to swell due to maximum speed of the rotating earth and the minimum Coriolis effect.

Chemical Composition

The chemical composition of this layer is basically uniform, with the important exception of water vapor. About 99% of the total water vapor in the atmosphere is present in the troposphere itself. The concentration of water vapor varies with latitudes. Equatorial region has maximum concentration of water vapor (up to 3%) and reduces towards polar region. At the same time water vapor is more at the surface and decreases with altitude. The troposphere primarily comprises of nitrogen (78%) and oxygen (21%) with very little amount of trace gases. Nearly 75% of the total mass of atmosphere is contained in troposphere and thus denser than the overlaying layers.

Temperature Characteristics

- Lapse rate: We usually feel cooler on the top of a mountain than in the valley below and chances of encountering snow or ice is greater in a higher hill than a hill with low altitude. Such thing happen as, troposphere, in general, shows a decreasing trend of air temperature with elevation. Air temperature usually declines from the earth's surface with increasing height up to about 11 km due to the fact that air near the surface gets heated up much more by earth's emitted long wave radiation energy than the air which is away from it. The rate at which the air temperature reduces with altitude is called the temperature lapse rate and average value of it is about 6.5 °C per km. This lapse rate is just an average value of day to day variations.

- Inversion: Sometimes, an increase in the air temperature with height may actually happen. Such a condition is known as temperature inversion and this prevents the vertical mixing of air. Temperature inversion plays an important role in air pollution where pollutants released at ground level get confined in the lower layers of atmosphere thus affecting the human and animal lives.

Tropopause-Isothermal Layer

Just above a certain height of the troposphere, the lapse rate becomes zero which means the air temperature (about - 60 °C) remains constant with altitude in that region i.e. an isothermal (equal temperature) layer. The bottom of this layer represents top of the troposphere and its upper limit as the starting point of the stratosphere. This transitional boundary differentiating the stratosphere from the troposphere is called the tropopause. It is typically higher over equatorial regions and lower in polar region. The height of tropopause generally increases in summer and decreases in winter at all latitudes. There are some 'breaks' in the tropopause where tropospheric air mixes with stratospheric air and vice versa. These breaks also indicate the position of high winds typically known as jet streams or 'river of air' having speeds exceeding 100 knots or 250 miles per hour.

- The troposphere harbors much of the weather which we experience on earth, beginning from wind to rain, thunderstorms and lightning, tornadoes, typhoons, hurricanes etc.
- Airplanes and jets flying at high altitudes must be adjusted with the lower pressure and density at that height.
- Maximum dust particles of the atmosphere are observed in the troposphere.
- Maximum air pressure is found at sea level and decreases as we go up. Similar is the trend for the amount of oxygen which is predominantly experienced by mountain climbers venturing to higher altitudes and needs artificial oxygen sources to survive.
- The troposphere facilitates temperature regulation and cloud formation. It has highest temperatures closer to its base; these warm temperatures help the troposphere retain water vapor, which later gets released in the form of precipitation.
- The exchange and movement of water between the earth and atmosphere is called the water cycle. The troposphere serves as the starting point for the earth's water cycle.

Stratosphere: Structure and Characteristics

The stratosphere extends up to an average height of 50 km from the earth surface. This layer is usually characterized by temperature inversion which keeps the vertical fluxes of the troposphere from dispersing into the stratosphere and thus generating a stratified layer.

Temperature Stratification and the Ozone Layer

Generally temperature rises in the stratosphere with height. The reason attributed to this is the presence of ozone gas. If ozone were absent, the air would become colder with height like in troposphere. Ozone is mainly confined in the lower stratosphere within 15–35 km and the thickness of ozone layer varies with seasons and latitudes with maximum at the equator and minimum at the poles. Stratospheric ozone constitutes about 90% of the total ozone present in our atmosphere with concentrations of about 2-8 ppm. The ozone (O_3) here absorbs high energy UV light and gets split into atomic oxygen (O) and diatomic oxygen (O_2). During formation of ozone by recombining these two forms of oxygen, heat is released into the stratosphere. Very low amount of UV light reaches to the lower part of stratosphere, thus atomic oxygen is not observed here and heat as the byproduct of ozone formation is not produced. This vertical stratification, where above layer becomes warmer and lower layer becomes cooler, makes the stratosphere dynamically stable. No convection and associated turbulence is found in this layer. The level of maximum ozone concentration is found at an altitude about 25 km over the middle latitudes. However, the stratospheric air temperature is maximum near 50 km due to the fact that at this specific height the air is less dense than at 25 km and the intense solar energy heats up the fewer molecules at 50 km which increases the temperature to a much greater degree. At the same time, the downward movement of heat energy is extremely slow due to the low air density which results in lower temperature in the stratum below.

The transition boundary layer from the top of the stratosphere to the beginning of mesosphere i.e. the stratopause is isothermal in nature. The air temperature of stratopause remains constant around 0 °C. Above stratopause, the temperature declines with height.

- Stratospheric ozone is vital for the survival of human beings and other life forms on earth, by absorbing UV radiation that would otherwise be fatal.
- Volcanic gases, dust, CFCs etc cause damage to thc ozone layer of the stratosphere.
- Stratosphere has extremely low amount of water vapor and thus we don't see any cloud there.
- Flying of airplanes in the stratosphere avoids turbulence, saves fuel burning and trip time.
- Sometimes research balloons with cameras are sent to record the weather information of the stratosphere.
- Stratosphere has no vertical winds: The temperature inversion in the stratosphere makes this layer very stable which does not allow significant vertical motion of winds. Thus vertical winds are nearly absent, in some cases may be up to a few centimeters per second. The incapability of the air to flow vertically is also the main reason why the Ozone depleting CFC takes so long to reach the height where the Sun's energy is sufficient enough to break them apart. This also implies that some of the ozone depleting substances will still be there a centuries later from now.
- Aviation and jet Streams in Stratosphere: Unlike troposphere, stratosphere is free from the harsh weather changes and is preferred by commercial aircrafts usually flying at altitudes of 9–12 km in the lower part of the stratosphere to adjust fuel burn. However, Jet liners face the problem of jet streams which are high velocity horizontal air currents and located near the tropopause. The location of the jet stream is particularly vital for aviation. Sometimes, jet streams plays important beneficial role in aviation by shorten the trip time and reducing the fuel consumption.

Mesosphere: Structure and Characteristics

The mesosphere, third layer in the atmosphere is situated 50 km from the earth's surface right above the stratopause and goes up to 85 km height (from the surface of the earth) i.e. till below the thermosphere.

The literal meaning of mesosphere is 'middle sphere' i.e. it exists in the middle of atmosphere. Mesosphere plays important role in protecting the planet from the meteors and asteroids falling to the earth by burning those with the gas particle present in the layer. The air is extremely thin here and the atmospheric pressure is very low, close to 1 mb indicating 99.9 percent of the atmospheric mass is found below this layer. The percentage of important gaseous components like nitrogen and oxygen in the mesosphere is similar which is seen at sea level. Due to very low density in this level, we cannot survive very long, as each breath will have far fewer oxygen molecules than at sea level.

Temperature

Air temperature usually drops when go up in the mesosphere as the air contains extremely low amount of ozone to absorb solar radiation. So, the upper part of the mesosphere loses more energy than it absorbs, which results in an energy deficit and cooling and the air temperature touches its lowest mean value, -90 °C. This is at the highest region in mesosphere and also the coldest place on earth.

Formation of Noctilucent Clouds

Unlike in troposphere, greenhouse gases, like carbon dioxide and methane behave differently in the mesosphere. Here, the carbon dioxide releases heat into the space from the mesosphere and cools down. At the same time, solar radiation disintegrates the methane gas. Reactions take place among CO_2, methane and ozone gas with water vapor present there under very low temperature of mesosphere, which leads to the formation of microscopic blue-white ice clouds popularly known as the noctilucent clouds. These beautiful clouds are clearly seen from the poles and during the sunset. A peculiar type of lightning is also observed in the mesosphere called 'Elves' or 'Sprites'.

Nightglow

If we see the clear sky at night properly, we can see some glowing layer due to unbound nonionized atoms of sodium having thickness of around 5 km. This sodium layer forms the nightglow which is the light emitted from the mesosphere and is located just below the mesopause.

Strong Winds

As temperature change is significant between the summer and winter poles, there are strong winds blowing from north to south in the lower mesosphere and in the upper mesosphere wind blows from east to west due to colder temperature there.

- Studies showed that the mesosphere burns roughly 40 tons of meteors falling towards earth each day otherwise they could reach the earth and make severe damage to its surface. Burning of meteors is seen as shooting stars in the night sky.

- Satellites can't stay in orbit in the mesosphere due to high atmospheric drag. The sounding rockets are used to study the mesosphere.
- The mesosphere is composed of oxygen, nitrogen and carbon dioxide and the air density is very low here.

Thermosphere: Structure and Characteristics

The "hot layer" above the mesosphere is the thermosphere. The transition boundary that divides the lower and colder mesosphere from the warmer thermosphere is the mesopause. The thermosphere derives its name from the Greek word 'thermos', meaning 'heat'. Starting from the surface, it is the fourth layer of the Earth's atmosphere. This also makes it the second farthest atmospheric layer from the Earth's surface extending vertically for a distance of about 250 to 550 miles; the only layer farther than it being the exosphere. The thermosphere is mainly comprised with helium, atomic oxygen and atomic nitrogen.

- The thermosphere can absorb the X-ray and the UV radiation coming from the sun.
- Solar activity increases the size of the thermosphere and makes it as one of the most dynamic layer. The only entity that can make it to this layer is a space shuttle. Excitingly, the orbit of the International Space Station (ISS) is placed in thermosphere.
- This layer exhibits some remarkable sky phenomena like the northern or southern lights.
- The charged particles of thermosphere greatly help in long distance radio communication.
- In the lower half of the thermosphere, the temperature increases very quickly with altitude. It is only after reaching a point that it becomes stable. In the upper thermosphere, temperature ranges from 500 °C to 2000 °C. The temperature here, is highly dependent on solar activity, and can reach 2000 °C or higher when solar activity is at its peak.
- Thin air also makes it difficult for the scientists to measure the temperature of this layer. Instead of measuring it directly, the scientists first measure the density of the air by measuring the drag on the satellites and use the same to determine the temperature. Other than density observations, scientists may also resort to direct satellite measurements to determine the temperature of this layer.

Exosphere: Structure and Characteristics

Above the earth's surface at the height of 500 km (top of the thermosphere), air is extremely thin with a mean free path (for gas molecules) of 10 km or more. Here, a number of fast-moving molecules or atoms have the capability of escaping the earth's gravitational pull. This region where atoms and molecules shoot off into space is known as the exosphere, which signifies the upper limit of our atmosphere. Due to ambiguity of clear cut demarcation between the exosphere and the space, it is hard to determine the height of this layer.

Layers of Atmosphere based on Electrical Properties

Ionosphere and its Sub Layers

The ionosphere is more appropriately described as an electrified region within the upper atmosphere rather than just a layer consisting of large concentrations of ions and free electrons. Ions are nothing but the charged atoms or molecules having deficient or extra electrons that are unable to absorb all of the transferred energy due to collision between colliding particles and radiation from the sun. Ionosphere stretches from 50 to 1,000 km and typically overlaps both the thermosphere and the exosphere. The lower levels of ionosphere have more number of ions compared to the upper levels. Hence, lower levels show the greater ionization and recombination as compared to the upper levels producing a highly unstable layer of the earth's atmosphere.

The ionization depends primarily on the amount of radiation received from the Sun and its activity. This is the reason that there are changes in the ionosphere on the diurnal and seasonal basis. The activity of the sun is associated with sunspot cycle; the higher the number of sunspots, more will be the solar output. Radiation received also varies with the position of earth in the revolutionary orbit and geographical location. There are some disturbances like solar flares or mechanisms which interrupt the ionosphere and decrease the ionization and also interact with its geomagnetic field. Accordingly, ionosphere has been divided into different sets of layers during day and night.

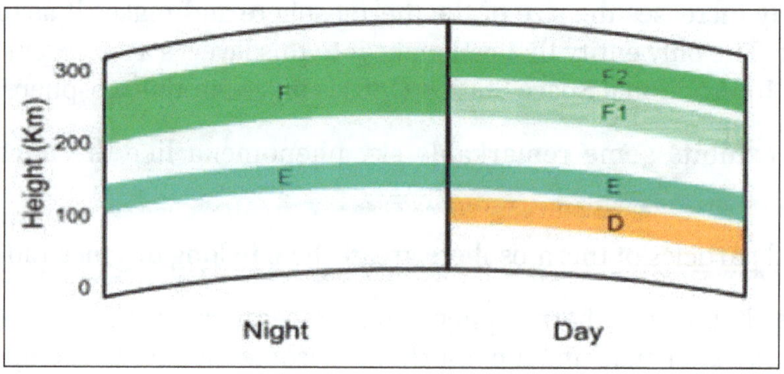

Sub layers of the Ionosphere.

D Layer

The D layer explains why the AM Radio gets disturbed during day time, but quite smooth in night time. We can see in the above figure that the D layer is situated between 60 km to 90 km above the earth surface and is the innermost layer. This layer shows the low net ionization effect, but loses huge amount of wave energy due to frequent collisions of the electrons. Thus, the high-frequency (HF) radio waves are least attenuated by the D layer but suffer loss of energy therein. The absorption is maximum about midday and minimum at night which results in the disappearance of distant AM radio waves in the daytime.

E Layer

The E layer is the intermediate layer, extending from 90 km to 120 km above the earth surface, with primary source of ionization being soft X-ray (1-10 nm) and far ultraviolet (UV) solar radiation causing ionization of molecular oxygen (O_2). This layer disappears in the night because primary source of ionization is no longer present. The practical value of this layer is that it reflects long radiowaves in the 1-3 megahertz (MHz) back to earth, which enables them to be received at a distance, rather than disappear into space. It is also known as Heavyside-Kennely layer.

F Layer

The F layer extends from about 200 km to more than 500 km above the surface of earth. The E-layer allows the penetration of short-radio waves, which continue until they reach the Appleton layer. Appleton layer reflects short-radio waves (which have penetrated the Heavyside-Kennely layer) back to earth. This is also supposed to be the region where polar auroras occur and where most of the meteors burn themselves out.

Ionosphere is also known as thermosphere because of the high temperatures prevailing there. The temperature near the upper boundary of the thermosphere may become higher than 1000-1500 °C. Along with temperature rise, sharp changes caused by the corpuscular and ultraviolet solar radiation are observed in it.

Some Important Phenomena of Ionosphere

Ionospheric Scintillation

Scintillation influences the phase and power of the propagating radio signal. Ionospheric scintillation is the fast alteration of radio waves produced by small scale structures in the ionosphere. Severe scintillation can stops a GPS receiver from getting the accurate signal and produces huge errors in position calculation. Low and high latitudes exhibit more prominent scintillation over mid-latitudes (e.g. the United States).

Aurora

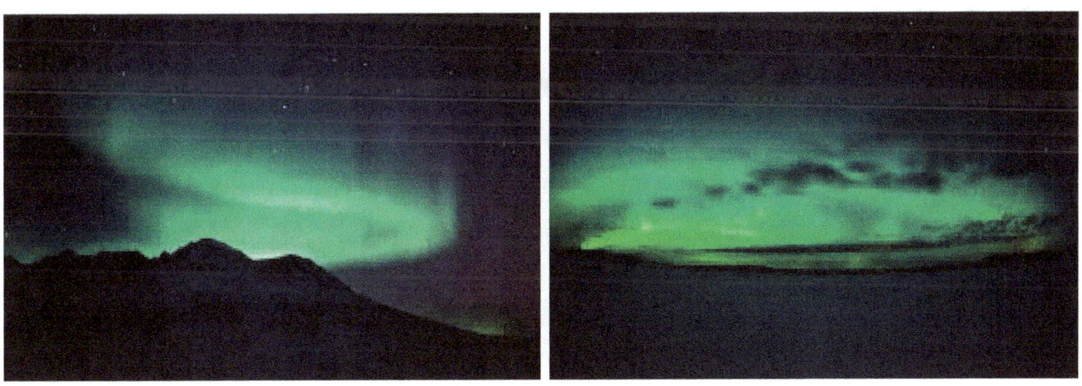

Aurora.

The aurora are illusive natural electro-magnetic phenomenon with luminous effect, which usually occurs near the northern (Aurora Borealis) and southern (Aurora Australis) poles in the night sky, best developed at a height of about 90 km. Such effects are probably the result of magnetic storms and of electrical discharges from the sun during periods of sun-spot activity, causing ionization of gases, though this is still a matter of research. The origin of aurora can be traced to solar flares which are huge eruptions on the surface of the sun that occur periodically (every 11 years). Most of the solar radiation returns back to the sun, but some ionized particles are ejected into space at very high speeds. These solar flares travel at the speed of roughly 1.5×10^6 m s^{-1} and enter the earth's atmosphere 30 hours after the occurrence of solar flares. Sine these particles are electrically charged, they are deflected towards the north and south poles by the magnetic fields of the earth. When these high-speed charged particles collide with the air molecules, they cause electrons to be freed. When the electrons recombine with air molecules, energy in the form of light is emitted. Occasionally the Aurora borealis is seen in England, but it is more common in northern Scotland. Figure below presents magnificent spectacles in northern Scandinavia and northern Canada.

Layers of Atmosphere based on Gaseous Composition

Based on chemical composition, the atmosphere may be divided into the following two regions:

- Homosphere: Below thermosphere, the composition of atmosphere is nearly uniform upto an altitude about 85 km (50 miles) due to turbulent mixing and hence, it is known as the homosphere. Though, the proportion of gases (nitrogen-78%, oxygen-21%, CO_2-0.03% etc.) remains almost unchanging in this well mixed region, the air density as well as air pressure declines significantly with increasing height.
- Heterosphere: The region starting from the base of the thermosphere and extending to the top of the atmosphere is often called the heterosphere. This means, the heterosphere is comprised of the thermosphere and the exosphere layers. Generally, heterosphere comprises less than 0.001% of the total mass of the earth's atmosphere. In this region, gases are not well mixed or stirred due to very infrequent collisions between atoms and molecules which results in accumulation of heavier gases like nitrogen and oxygen at the bottom and lighter ones like helium and hydrogen at the top of the layer by the process of diffusion.

Atmospheric Models

Four of the most frequently used words in aerodynamics:

- Pressure (p),
- Density (ρ),
- Temperature (T),
- Flow Velocity (V).

Pressure

The pressure is defined at a point in the fluid or a point on a solid surface. The pressure can vary from one point to another. "Pressure is the normal force per unit area exerted on a surface due to the time rate of change of momentum of the gas molecules impacting on (or crossing) that surface".

$$\rho = \lim_{dA \to 0} \left(\frac{dF}{dA} \right) N/m^2$$

- dA = Elemental Area at B (m^2).
- dF = Force on 1 side of dA due to pressure (N).

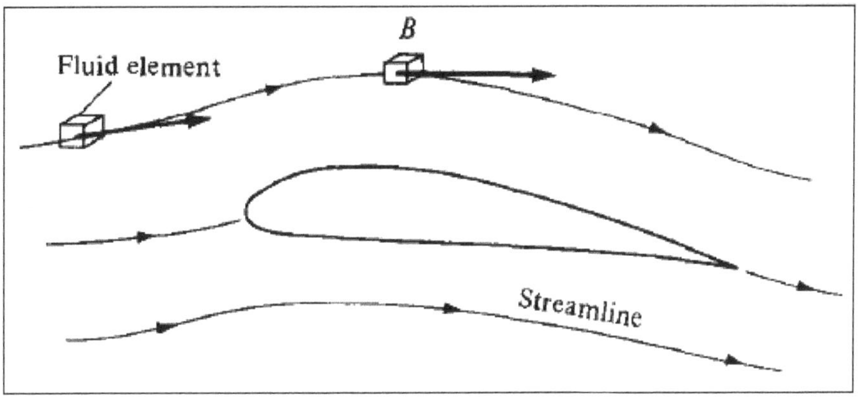

Density

The density of a material is a measure of the amount of material contained in a given volume:

$$\rho = \lim_{dv \to o}\left(\frac{dm}{dv}\right) kg/m^3$$

- dv = Elemental volume around B (m^3).
- dm = Mass of the fluid inside dv (kg).

Temperature

"The temperature T of a gas is directly proportional to the average kinetic energy of the molecules of the fluid. In fact, if KE is the mean molecular kinetic energy, then temperature is given by":

$$KE = \frac{3}{2}kT$$

Where k is the Boltzmann constant ($k = 1.38 \times 10^{-23}$ J/°K).

Temperature is "a point property, which can vary from point to point in the gas" & has "an important role in high-speed aerodynamics".

Flow Velocity

A velocity is a vector value & as such must contain both scalar value & direction. In a flowing fluid at each region (or point) in the fluid there is not necessarily the same velocity. Hence this is also a point property which can vary from point to point in the flow. "The velocity of a flowing gas at any fixed point B in space is the velocity of an infinitesimally small fluid element as it sweeps through B."

Viscosity of oil > Viscosity of air

Sutherland's Law

$$\frac{\mu}{\mu_o} = \left(\frac{T}{T_o}\right)^{3/2} \frac{T_o + 110}{T + 110}$$

Where:

- o is reference data.
- $T_o = 288.16$ °K.
- $\mu_o = 1.7894 \times 10^{-5}$ kg/ms.

Dynamic Viscosity

$$v = \frac{\mu}{\rho}$$

Newtonian Fluids

For many simple fluids, such as air and water, μ is a thermodynamic property which depends only on Temperature & Pressure, but not on the shear rate $\left(\tau = \mu \frac{\partial u}{\partial y}\right)$.

Newtonian ≠ Non Newtonian Fluids

Equation of State

This may also be referred to as the Ideal Gas Law or the Perfect Gas Law.

$$p = \rho RT$$

R = Specific Gas Constant = 287.05287 J/kg.°K

$$R = c_p - c_v \;\; \& \;\; \gamma = \frac{c_p}{c_v}$$

Where c_p & c_v are the specific heats.

The Hydrostatic Equation

For a fluid at rest, the pressure is constant over any horizontal surface but decreases with altitude.

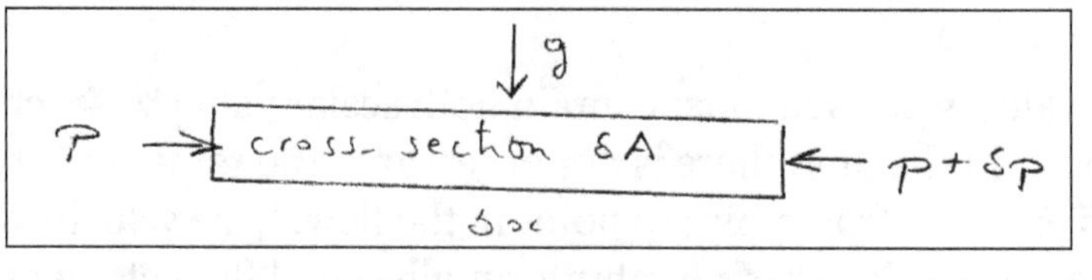

If the pressure is assumed to change from 'p' at one end of the cylinder to 'p + δp' at the other end, then the net force acting on the cylinder is:

$$p.\delta A - (p + \delta)\delta A = -\delta p.\delta A$$

This net force must be o unless the fluid cylinder is accelerating to the right ($\delta p < 0$) or to the left ($\delta p > 0$). Hence for a fluid at rest, $\delta p = 0$ & the pressure is constant along any horizontal line (or surface).

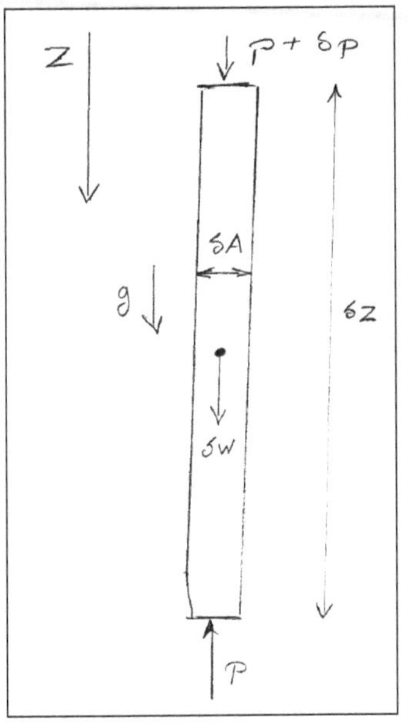

The net downwards force acting on the fluid cylinder is:

$$-p.\delta A + \delta W + (p + \delta p)\delta A = 0$$

Where:

$$\delta W = \rho(\delta z.\delta A)g$$

$$\therefore -p.\delta A + \rho(\delta z.\delta A)g + p.\delta A + \delta p.\delta A = 0$$
$$\therefore \rho.\delta z.\delta A.g + \delta p.\delta A = 0$$

$$\delta p = -\rho.\delta z.g$$

or,

$$\frac{dp}{dz} = -\rho.g$$

An Energy Balance Equation

- Adiabatic Process: One in which no heat is added to or taken away from the system.
- Isothermal Process: One in which the temperature remains constant.
- Reversible Process: One in which no dissipative phenomena occur, that is, where the effects of viscosity, thermal conductivity & mass diffusion are absent.
- Isentropic Process: One that is both Adiabatic & Reversible. If there is no heat transfer or friction in the process.

Isothermal Case

If, $T = T_o$

Then the equation of state becomes:

$$p = \rho R T_o$$

$$\Rightarrow \rho = \frac{p}{RT_o}$$

Substitute above equation into $\frac{dp}{dz} = -\rho.g$:

$$\Rightarrow \frac{dp}{dz} = -\frac{p}{RT_o} \times g$$

$$\frac{dp}{p} = -\frac{g}{RT_o} \times dz$$

$$\rho = p_o, e^{\left(-g/RT_o\right)z}$$

Where p_o occurs at z = 0.

Substitute above equation $\Rightarrow \rho = p_o, e^{\left(-g/RT_o\right)z}$ into $\Rightarrow \rho = \frac{p}{RT_o}$:

$$\rho = \frac{\rho_o}{RT_o}, e^{(-g/RT_o)z}$$

$$\rho = \rho_o, e^{(-g/RT_o)z}$$

Where ρ_o occurs at z = 0.

Adiabatic Process

$$\delta q = 0$$

For an isentropic process, ds = 0, where s is the entropy of the system.

Entropy and the second law of thermodynamics:

$$ds = \frac{\delta q}{T}$$

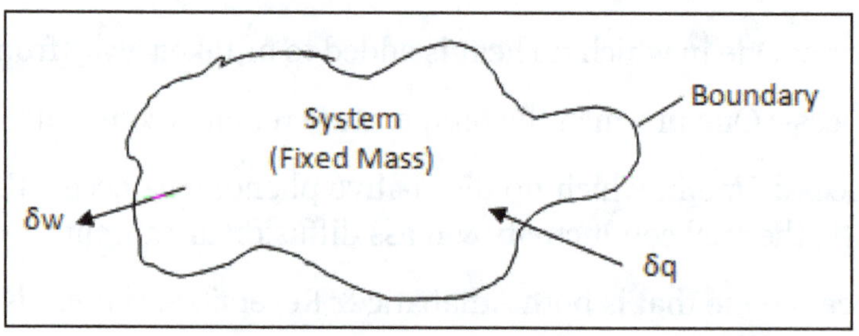

Thermodynamic System

First Law of Thermodynamics

The change in energy in a system is:

$$de = \delta w + \delta q$$

For a reversible process: $\delta w = -pdv$

$$de = \delta q - pdv$$

Where dv is an elemental change in the volume due to a displacement of the boundary of the system.

By deriving $ds = \frac{\delta q}{T}$ it can be shown that $\delta q = Tds$

$$\therefore de = Tds - pdv$$

$$\Rightarrow Tds = de + p.dv$$

The enthalpy is defined by:

$$h = e + pv$$

$$\therefore dh = de + d(pv) = de + p.dv + v.dp$$
$$\therefore de + p.dv = dh - v.dp$$

Substitute this into $\Rightarrow Tds = de + p.dv$:

$$\Rightarrow Tds = dh - v.dp$$

For a Perfect Gas

$$de = c_v.dT \qquad \& \qquad dh = c_p.dT$$

Therefore by substituting into $\Rightarrow Tds = de + p.dv$ and $\Rightarrow Tds = dh - v.dp$:

$$\Rightarrow ds = c_v \frac{dT}{T} + \frac{p.dv}{T}$$
$$\Rightarrow ds = c_p \frac{dT}{T} - \frac{v.dp}{T}$$

Using the Equation of State:

$$\frac{p}{\rho} = RT \quad \text{or} \quad pv = RT$$

$$\therefore v = \frac{RT}{p} \quad \text{or} \quad p = \frac{RT}{v}$$

Substitute these into above equations:

$$ds = c_v \frac{dT}{T} + R\frac{dv}{v} \Leftrightarrow ds = c_p \frac{dT}{T} - R\frac{dp}{p}$$

$$s_2 - s_1 = c_v . \ln\left(\frac{T_2}{T_1}\right) + R.\ln\left(\frac{v_2}{v_1}\right) \Leftrightarrow s_2 - s_1 = c_p . \ln\left(\frac{T_2}{T_1}\right) + R.\ln\left(\frac{p_2}{p_1}\right)$$

For an isentropic process, δs = 0, this can be substituted into above equations:

$$0 = c_v . \ln\left(\frac{T_2}{T_1}\right) + R.\ln\left(\frac{v_2}{v_1}\right) \ \& \ 0 = c_p . \ln\left(\frac{T_2}{T_1}\right) + R.\ln\left(\frac{p_2}{p_1}\right)$$

Therefore:

$$\ln\left(\frac{v_2}{v_1}\right) = \frac{c_v}{R} . \ln\left(\frac{T_2}{T_1}\right) \quad \& \quad \ln\left(\frac{p_2}{p_1}\right) = \frac{c_p}{R} . \ln\left(\frac{T_2}{T_1}\right)$$

Therefore:

$$\frac{v_2}{v_1} = \left(\frac{T_2}{T_1}\right)^{\frac{c_v}{R}} \Leftrightarrow \frac{p_2}{p_1} = \left(\frac{T_2}{T_1}\right)^{\frac{c_p}{R}}$$

Where:

$$R = c_p - c_v \ \& \ \gamma = \frac{c_p}{c_v}$$

Therefore:

$$c_v = \frac{R}{\gamma - 1} \Leftrightarrow c_p = \frac{\gamma R}{\gamma - 1}$$

Substitute above into $\frac{v_2}{v_1} = \left(\frac{T_2}{T_1}\right)^{\frac{c_v}{R}} \Leftrightarrow \frac{p_2}{p_1} = \left(\frac{T_2}{T_1}\right)^{\frac{c_p}{R}}$:

$$\frac{v_2}{v_1} = \left(\frac{T_2}{T_1}\right)^{\frac{-1}{(\gamma-1)}} \Leftrightarrow \frac{p_2}{p_1} = \left(\frac{T_2}{T_1}\right)^{\frac{\gamma}{(\gamma-1)}}$$

Where:

$$v = \frac{1}{\rho} \quad \therefore \frac{v_2}{v_1} = \frac{\rho_1}{\rho_2}$$

Substitute this into $\frac{v_2}{v_1} = \left(\frac{T_2}{T_1}\right)^{\frac{-1}{(\gamma-1)}} \Leftrightarrow \frac{p_2}{p_1} = \left(\frac{T_2}{T_1}\right)^{\frac{\gamma}{(\gamma-1)}}$:

$$\frac{\rho_1}{\rho_2}=\left(\frac{T_2}{T_1}\right)^{\frac{-1}{(\gamma-1)}}$$

Therefore:

$$\frac{p_2}{p_1}=\left(\frac{\rho_2}{\rho_1}\right)^{\gamma}=\left(\frac{T_2}{T_1}\right)^{\frac{\gamma}{(\gamma-1)}}$$

An adiabatic process is characterised by:

$$\frac{p}{\rho^{\gamma}}=cst$$

Constant Lapse Rate

In the troposphere, the temperature decreases with altitude z, according to the relation:

$$T=T_o-\gamma z$$

Where T_o = 288.16 °K at z = 0 & γ is the lapse rate.

$$\therefore dT=-\gamma.dz$$

$$\Rightarrow \gamma=-\frac{dT}{dz}$$

In isothermal conditions $T = T_o$ = Constant. Therefore γ = 0

Using the hydrostatic equation derived from $\Rightarrow \rho=\frac{p}{RT_o}$:

$$\Rightarrow \frac{dp}{p}=\frac{-g}{R(T_o-\gamma z)}dz$$

$$\therefore \frac{dp}{p}=\frac{\gamma g}{-\gamma R}\frac{dz}{(T_o-\gamma z)}$$

$$\therefore \int\frac{dp}{p}=\int\frac{g}{\gamma R}\left(\frac{-\gamma.dz}{T_o-\gamma z}\right)$$

$$\Rightarrow \frac{p}{p_o}=\left(\frac{T}{T_o}\right)^{g/\gamma R}$$

Using the Equation of State:

$$\Rightarrow \frac{\rho}{\rho_o}=\left(\frac{T}{T_o}\right)^{(g/\gamma R)}$$

Fluid Dynamics

Fluid dynamics is "the branch of applied science that is concerned with the movement of liquids and gases," according to the American Heritage Dictionary. Fluid dynamics is one of two branches of fluid mechanics, which is the study of fluids and how forces affect them. (The other branch is fluid statics, which deals with fluids at rest).

Scientists across several fields study fluid dynamics. Fluid dynamics provides methods for studying the evolution of stars, ocean currents, weather patterns, plate tectonics and even blood circulation. Some important technological applications of fluid dynamics include rocket engines, wind turbines, oil pipelines and air conditioning systems.

Flow

The movement of liquids and gases is generally referred to as "flow," a concept that describes how fluids behave and how they interact with their surrounding environment — for example, water moving through a channel or pipe, or over a surface. Flow can be either steady or unsteady. In his lecture notes, "Lectures in Elementary Fluid Dynamics" J. M. McDonough, a professor of engineering at the University of Kentucky, writes, "If all properties of a flow are independent of time, then the flow is steady; otherwise, it is unsteady." That is, steady flows do not change over time. An example of steady flow would be water flowing through a pipe at a constant rate. On the other hand, a flood or water pouring from an old-fashioned hand pump are examples of unsteady flow.

Flow can also be either laminar or turbulent. Laminar flows are smoother, while turbulent flows are more chaotic. One important factor in determining the state of a fluid's flow is its viscosity, or thickness, where higher viscosity increases the tendency of the flow to be laminar. Patrick McMurtry, an engineering professor at the University of Utah, describes the difference in his online class notes, "Observations about Turbulent Flows", stating, "By laminar flow we are generally referring to a smooth, steady fluid motion, in which any induced perturbations are damped out due to the relatively strong viscous forces. In turbulent flows, other forces may be acting the counteract the action of viscosity."

Laminar flow is desirable in many situations, such as in drainage systems or airplane wings, because it is more efficient and less energy is lost. Turbulent flow can be useful for causing different fluids to mix together or for equalizing temperature. According to McDonough, most flows of interest are turbulent; however, such flows can be very difficult to predict in detail, and distinguishing between these two types of flow is largely intuitive.

An important factor in fluid flow is the fluid's Reynolds number (Re), which is named after 19th century scientist Osborne Reynolds, although it was first described in 1851 by physicist George Gabriel Stokes. McDonough gives the definition of Re as, "the ratio of inertial to viscous forces." The inertial force is the fluid's resistance to change of motion, and the viscous force is the amount of friction due to the viscosity or thickness of the fluid. Note that Re is not only a property of the fluid; it also includes the conditions of its flow such as its speed and the size and shape of the conduit or any obstructions.

At low Re, the flow tends to be smooth, or laminar, while at high Re, the flow tends to be turbulent, forming eddies and vortices. Re can be used to predict how a gas or liquid will flow around an obstacle in a stream, such as water around a bridge piling or wind over an aircraft wing. The number can also be used to predict the speed at which flow transitions from laminar to turbulent.

Liquid Flow

The study of liquid flow is called hydrodynamics. While liquids include all sorts of substances, such as oil and chemical solutions, by far the most common liquid is water, and most applications for hydrodynamics involve managing the flow of this liquid. That includes flood control, operation of city water and sewer systems, and management of navigable waterways.

Gas Flow

The flow of gas has many similarities to the flow of liquid, but it also has some important differences. First, gas is compressible, whereas liquids are generally considered to be incompressible. "If the density of the fluid changes appreciably throughout the flow field, the flow may be treated as a compressible flow." Otherwise, the fluid is considered to be incompressible. Second, gas flow is hardly affected by gravity.

The gas most commonly encountered in everyday life is air; therefore, scientists have paid much attention to its flow conditions. Wind causes air to move around buildings and other structures, and it can also be made to move by pumps and fans.

One area of particular interest is the movement of objects through the atmosphere. This branch of fluid dynamics is called aerodynamics, which is "the dynamics of bodies moving relative to gases, especially the interaction of moving objects with the atmosphere," according to the American Heritage Dictionary. Problems in this field involve reducing drag on automobile bodies, designing more efficient aircraft and wind turbines, and studying how birds and insects fly.

Bernoulli's Principle

Generally, fluid moving at a higher speed has lower pressure than fluid moving at a lower speed. This phenomenon was first described by Daniel Bernoulli in 1738 in his book "Hydrodynamica," and is commonly known as Bernoulli's principle. It can be applied to measure the speed of a liquid or gas moving in a pipe or channel or over a surface.

This principle is also responsible for lift in an aircraft wing, which is why airplanes can fly. Because the wing is flat on the bottom and curved on the top, the air has to travel a greater distance along the top surface than along the bottom. To do this, it must go faster over the top, causing its pressure to decrease. This makes the higher-pressure air on the bottom lift up on the wing.

Problems in Fluid Dynamics

Scientists often try to visualize flow using figures called streamlines, streaklines and pathlines. McDonough defines a streamline as "a continuous line within a fluid such that the tangent at each point is the direction of the velocity vector at that point." In other words, a streamline shows the direction of the flow at any particular point in the flow. A streakline, according to McDonough, is "the locus [location] of all fluid elements that have previously passed through a given point." A pathline (or particle path), he writes, is "the trajectory of an individual element of fluid." If the flow does not change over time, the pathline will be the same as the streamline. However, in the case of turbulent or unsteady flow, these lines can be quite different.

Most problems in fluid dynamics are too complex to be solved by direct calculation. In these cases, problems must be solved by numeric methods using computer simulations. This area of study is called numerical or computational fluid dynamics (CFD), which Southard defines as "a branch of computer-based science that provides numerical predictions of fluid flows." However, because turbulent flow tends to be nonlinear and chaotic, particular care must be taken in setting up the rules and initial conditions for these simulations. Small changes at the beginning can result in large differences in the results. The accuracy of simulations can be improved by dividing the volume into smaller regions and using smaller time steps, but this increases computing time. For this reason, CFD should advance as computing power increases.

Equilibrium of a Fluid Element

If the sum of the external forces acting on a fluid element is zero, the fluid will be either at rest or moving as a solid body - in either case, we say the fluid element is in equilibrium. For fluids in equilibrium the only internal stresses acting will be normal forces, since the shear stresses depend on velocity gradients, and all such gradients, by the definition of equilibrium, are zero. If one then carries out a balance between the normal surface stresses and the body forces, assumed proportional to volume or mass, such as gravity, acting on an elementary prismatic fluid volume, the resulting equilibrium equations, after shrinking the volume to zero, show that the normal stresses at a point are the same in all directions, and since they are known to be negative, this common value is denoted by -p, p being the pressure.

Hydrostatic Pressure

If we carry out equilibrium of forces on an elementary volume element dxdydz, the forces being pressures acting on the faces of the element and gravity acting in the -z, direction, we obtain:

$$\frac{\partial p}{\partial x} = \frac{\partial p}{\partial y} = 0, \quad \text{and} \quad \frac{\partial p}{\partial z} = -\rho g = -\gamma$$

The first two of these imply that the pressure is the same in all directions at the same vertical height in a gravitational field. The third, where γ is the specific weight, shows that the pressure increases with depth in a gravitational field, the variation depending on $\rho(z)$. For homogeneous fluids, for which ρ = constant, this last equation can be integrated immediately, yielding:

$$p_2 - p_1 = -\rho g(z_2 - z_1) = -\rho g(h_2 - h_1)$$

or

$$p_2 + \rho g h_2 = p_1 + \rho g h_1 = \text{constant}$$

Where h denotes the elevation. These are the equations for the hydrostatic pressure distribution. When applied to problems where a liquid, such as the ocean, lies below the atmosphere, with a constant pressure p_{atm}, h is usually measured from the ocean/atmosphere interface and p at any distance h below this interface differs from p_{atm} by an amount:

$$p - p_{atm} = \rho g h$$

Pressures may be given either as absolute pressure, pressure measured relative to absolute vacuum, or gauge pressure, pressure measured relative to atmospheric pressure.

Manometry

The hydrostatic pressure variation may be employed to measure pressure differences in terms of heights of liquid columns - such devices are called manometers and are commonly used in wind tunnels and a host of other applications and devices. Consider, for example the U-tube manometer shown in Figure below filled with liquid of specific weight γ, the left leg open to the atmosphere and the right to the region whose pressure p is to be determined. In terms of the quantities shown in the figure, in the left leg:

$$p_o - \rho g h_2 = p_{atm}$$

and in the right leg:

$$p_o - \rho g h_1 = p$$

the difference being:

$$p - p_{atm} = -\rho g(h_1 - h_2) = -\rho g d = -\gamma d$$

and determining *p* in terms of the height difference $d = h_1 - h_2$ between the levels of the fluid in the two legs of the manometer.

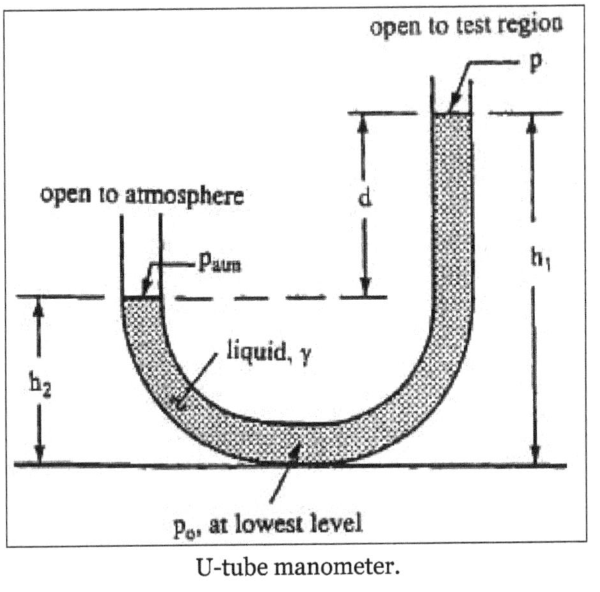

U-tube manometer.

Hydrostatic Forces on Submerged Objects

We now consider the force acting on a submerged object due to the hydrostatic pressure. This is given by:

$$F = \iint p\,dA = \iint p \cdot n\,dA = \iint \rho g h\,dA + p_o \iint dA$$

where h is the variable vertical depth of the element dA and p_o is the pressure at the surface. In turn we consider plane and nonplanar surfaces.

Forces on Plane Surfaces

Consider the planar surface A at an angle θ to a free surface shown in figure below. The force on one side of the planar surface, from Equation above, is:

$$F = \rho g n \iint_A h\,dA + p_o A n$$

But $h = y \sin\theta$, so:

$$\iint_A h\,dA = \sin\theta \iint_A y\,dA = y_c A \sin\theta = h_c A$$

where the subscript c indicates the distance measured to the centroid of the area A. Thus, the total force (on one side) is:

$$F = \gamma h_c A + p_o A$$

Thus, the magnitude of the force is independent of the angle θ, and is equal to the pressure at the centroid, $\gamma h_c + p_o$, times the area. If we use gauge pressure, the term $p_o A$ in Equation above can be dropped.

Since p is not evenly distributed over A, but varies with depth, F does not act through the centroid. The point action of F, called the center of pressure, can be determined by considering moments in Figure below. The moment of the hydrostatic force acting on the elementary area dA about the axis perpendicular to the page passing through the point o on the free surface is:

$$\gamma dF = y(\gamma\, y \sin\theta\, dA) = \gamma\, y^2 \sin\theta\, dA$$

so if y_{cp} denotes the distance to the center of pressure,

$$y_{cp} F = \gamma \sin\theta \iint y^2 dA = \gamma \sin\theta I_x$$

where I_x is the moment of inertia of the plane area with respect to the axis formed by the intersection of the plane containing the planar surface and the free surface (say ox). Dividing by $F = \gamma h_c A = \gamma\, y_c \sin\theta A$ gives:

$$y_{cp} = \frac{I_x}{y_c A}$$

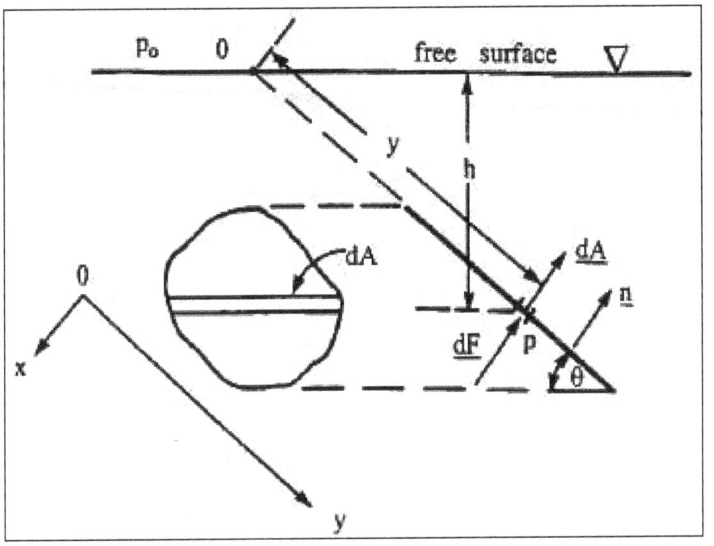

Hydrostatic force on a plane surface.

By using the parallel axis theorem $I_x = I_{xc} + Ay_c^2$ where I_{xc} is the moment of inertia with respect to an axis parallel to o*x* passing through the centroid, Equation above becomes:

$$y_{cp} = y_c + \frac{I_{xc}}{y_c A}$$

which shows that, in general, the center of pressure lies below the centroid. Similarly, we find x_{cp} by taking moments about the *y* axis, specifically,

$$x_{cp} F = \gamma \sin\theta \iint xy \, dA = \gamma \sin\theta I_{xy}$$

or,

$$x_{cp} = \frac{I_{xy}}{y_c A}$$

where I_{xy} is the product of inertia with respect to the *x* and *y* axes. Again, the parallel axis theorem $I_{xy} = I_{xyc} + Ax_c y_c$, where the subscript c denotes the value at the centroid, allows Equation above to be written:

$$x_{cp} = x_c + \frac{I_{xyc}}{y_c A}$$

This completes the determination of the center of pressure (x_{cp}, y_{cp}). Note that if the submerged area is symmetrical with respect to an axis passing through the centroid and parallel to either the x or y axes that $I_{xyc} = 0$ and $x_{cp} = x_c$; also that as y_c increases, $y_{cp} \rightarrow y_c$.

Centroidal moments of inertia and centroidal coordinates for some common areas are shown in Figure below.

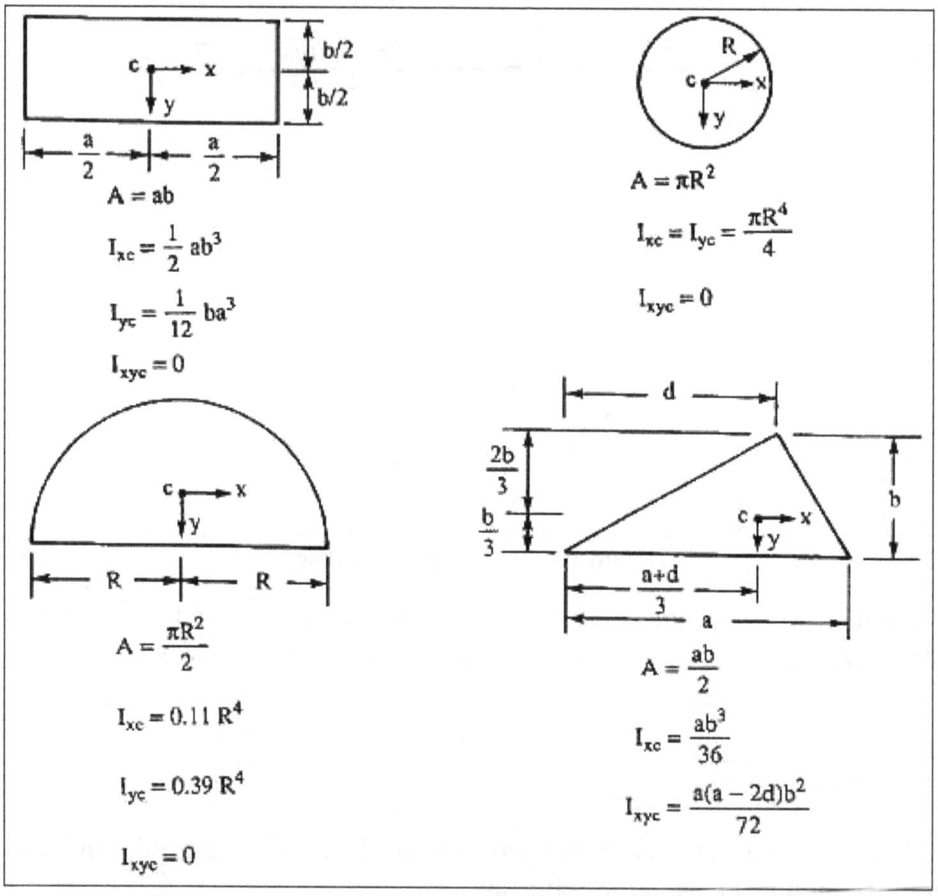

Centroidal moments of inertia and coordinates for some common areas.

Forces on Curved Surfaces

On a curved surface the forces on individual elements of area differ in direction so a simple summation of them is not generally possible, and the most convenient approach to calculating the pressure force on the surface is by separating it into its horizontal and vertical components. A free-body diagram of the forces acting on the volume of fluid lying above a curved surface together with the conditions of static equilibrium of such a column leads to the results that:

- The horizontal components of force on a curved submerged surface are equal to the forces exerted on the planar areas formed by the projections of the curved surface onto vertical planes normal to these components, the lines of action of these forces calculated as described earlier for planar surfaces.
- The vertical component of force on a curved submerged surface is equal in magnitude to the weight of the entire column of fluid lying above the curved surface, and acts through the center of mass of this volume of fluid.

Since the three components of force, two horizontal and one vertical, calculated as above, need not meet at a single point, there is, in general, no single resultant force. They can, however, be combined into a single force at any arbitrary point of application together with a moment about that point.

Hydrostatic Forces in Layered Fluids

All of the above results which employ the linear hydrostatic variation of pressure are valid only for homogeneous fluids. If the fluid is heterogeneous, consisting of individual layers each of constant density, then the pressure varies linearly with a different slope in each layer and the preceding analyses must be remedied by computing and summing the separate contributions to the forces and moments.

Buoyancy

The same principles to compute hydrostatic forces can be used to calculate the net pressure force acting on completely submerged or floating bodies. These laws of buoyancy, the principles of Archimedes, are that:

- A completely submerged body experiences a vertical upward force equal to the weight of the displaced fluid.
- A floating or partially submerged body displaces its own weight in the fluid in which it floats (i.e., the vertical upward force is equal to the body weight).

The line of action of the buoyancy force in both (1) and (2) passes through the centroid of the displaced volume of fluid; this point is called the center of buoyancy. (This point need not correspond to the center of mass of the body, which could have non-uniform density. It has been assumed that the displaced fluid has a constant g. If this is not the case, such as in a layered fluid, the magnitude of the buoyant force is still equal to the weight of the displaced fluid, but the line of action of this force passes through the center of gravity of the displaced volume, not the centroid). If a body has a weight exactly equal to that of the volume of fluid it displaces, it is said to be neutrally buoyant and will remain at rest at any point where it is immersed in a (homogeneous) fluid.

Stability of Submerged and Floating Bodies

Submerged Body

A body is said to be in stable equilibrium if, when given a slight displacement from the equilibrium position, the forces thereby created tend to restore it back to its original position. The forces acting on a submerged body are the buoyancy force, FB, acting through the center of buoyancy, denoted by CB, and the weight of the body, W, acting through the center of gravity denoted by CG.

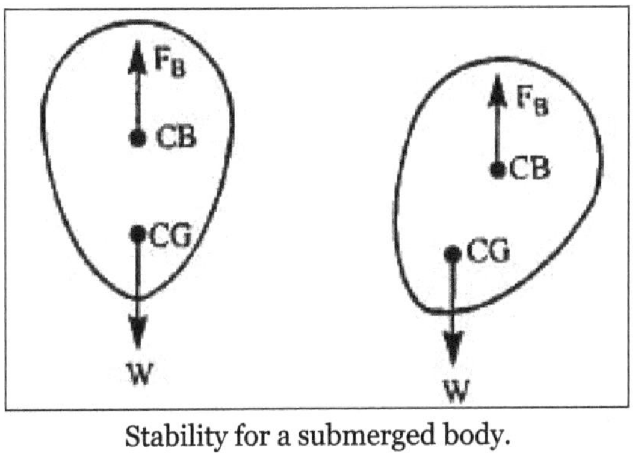

Stability for a submerged body.

We see from Figure below that if the CB lies above the CG a rotation from the equilibrium position creates a restoring couple which will rotate the body back to its original position - thus, this is a stable equilibrium situation. The reader will readily verify that when the CB lies below the CG, the couple that results from a rotation from the vertical increases the displacement from the equilibrium position - thus, this is an unstable equilibrium situation.

Partially Submerged Body

The stability problem is more complicated for floating bodies because as the body rotates the location of the center of buoyancy may change. To determine stability in these problems requires that we determine the location of the metacenter. This is done for a symmetric body by tilting the body through a small angle from its equilibrium position and calculating the new location of the center of buoyancy CB'; the point of intersection of a vertical line drawn upward from CB' with the line of symmetry of the floating body is the metacenter, denoted by M in Figure below, and it is independent of $\Delta\theta$ for small angles. If M lie $\Delta\theta$ s above the CG of the body, we see from Figure below that rotation of the body leads to a restoring couple, whereas M lying below the CG leads to a couple which will increase the displacement.

Thus, the stability of the equilibrium depends on whether M lies above or below the CG. The directed distance from CG to M is called the metacentric height, so equivalently the equilibrium is stable if this vector is positive and unstable if it is negative; stability increases as the metacentric height increases. For geometrically complex bodies, such as ships, the computation of the metacenter can be quite complicated.

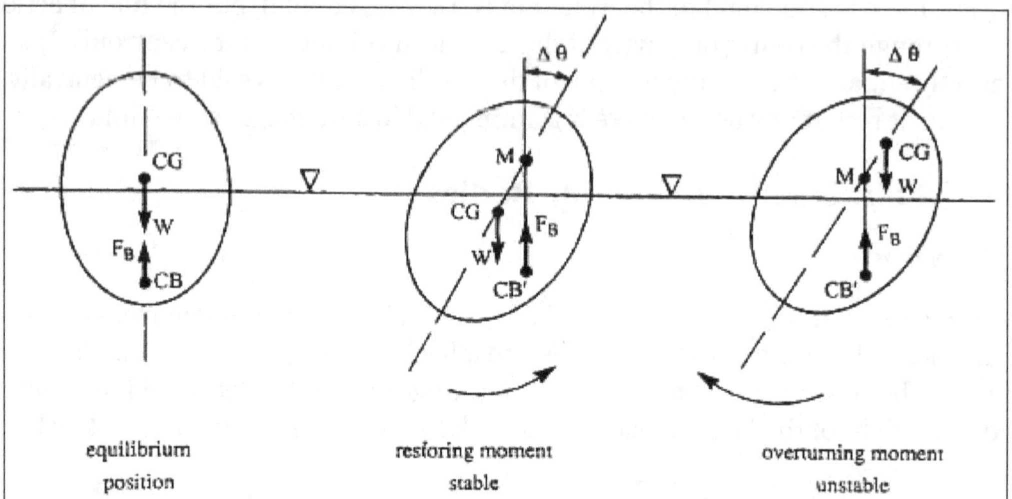

Stability for a partially submerged body.

Computational Fluid Dynamics

Computational Fluid Dynamics (CFD) is the analysis of fluid flows using numerical solution methods. Using CFD, you are able to analyze complex problems involving fluid-fluid, fluid-solid or fluid-gas interaction. Engineering fields where CFD analyses are frequently used are for example aerodynamics and hydrodynamics, where quantities such as lift and drag or field properties as pressures and velocities are obtained. Fluid dynamics is involved with physical laws in the form of partial differential equations. Sophisticated CFD solvers transform these laws into algebraical equations and are able to efficiently solve these equations numerically.

Some CFD simulation software provides a complete environment for performing analysis such as heat transfer by combining all the phases of analysis in one single package: from solid modeling, to problem set up, running, results visualization, validating the design, and reporting. Figure illustrates some of the important characteristics of CFD software that enables it to be useful to design engineers and not just specialists.

MCAD System Embedded

All the designer needs is knowledge of the MCAD system and the physics of the product to use next-generation CFD. After installation, all the menus and commands necessary to run a full CFD flow analysis are created in the MCAD package's menu system. This CFD embedded within the MCAD system makes it extremely easy to use. The design engineer is already familiar with the menu structures. The CFD analysis tool uses the same GUI structures so the engineer does not have to learn a completely new interface.

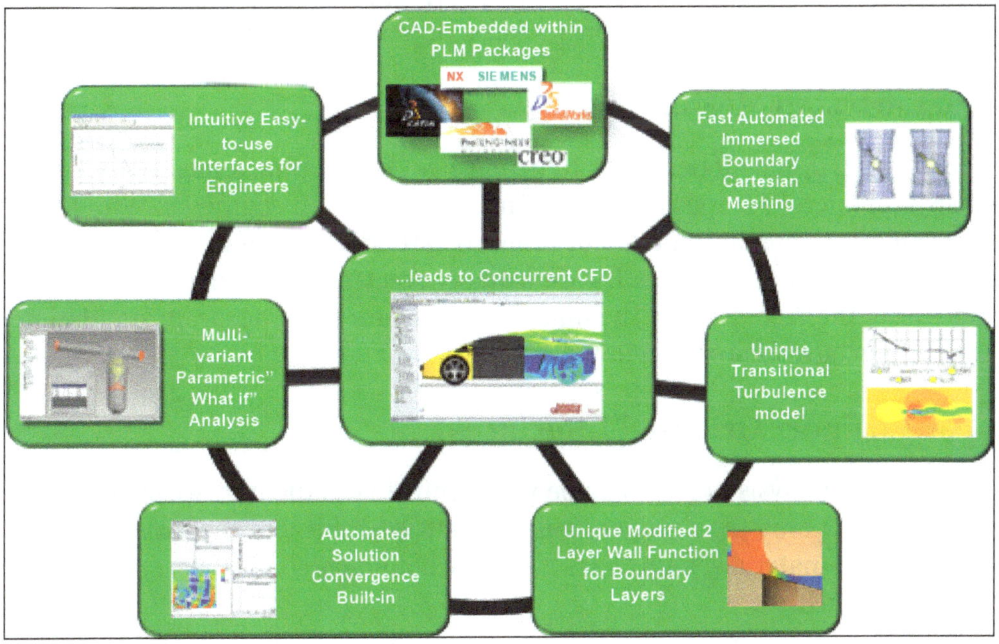

Enabling design engineers as well as specialists to perform CFD analysis requires technologies that automate difficult processes like meshing, integration with the MCAD system, guidance through the setup with wizards, etc.

The starting point of any heat transfer and fluid flow analysis is to define the overall boundary conditions of the problem. A wizard is available to direct the setup, including the selection of models. This lets a designer take advantage of existing MCAD models for analysis, without having to export or import additional data, saving significant amount of time and effort. The embedded CFD software can use newly created or existing 3D CAD geometry and solid model information to simulate designs in real-world conditions.

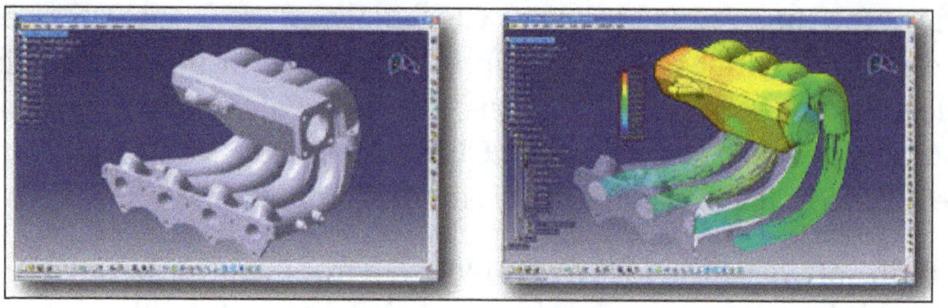

This example shows the FloEFD analysis software as embedded into the CATIAV5 MCAD system—same look/feel and GUI.

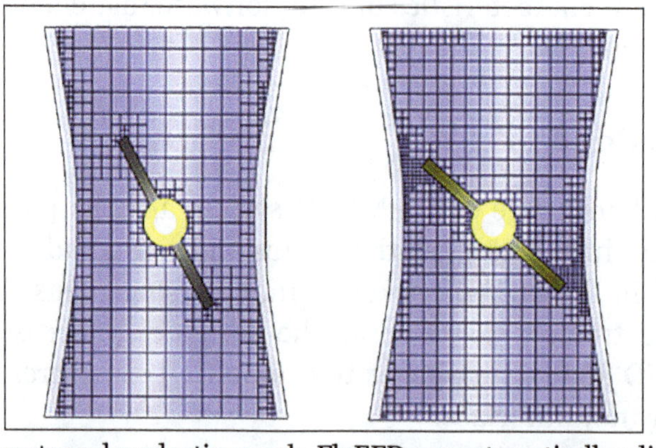

Using a rectangular adaptive mesh, FloEFD can automatically adjust cell size to deliver better resolution anywhere it is needed.

Automated Cartesian Meshing

Once a model is created, the model needs to be meshed. Developing a mesh is one of those skills that previously separated CFD specialists from mechanical design engineers. With FloEFD, meshes are created automatically in a matter of minutes rather than requiring hours of tedious proportioning of regions and cells. The software actually creates an adaptive mesh that reduces the cell size (increasing the resolution of the analysis) to ensure more accurate simulation results in complex areas of the model, as shown in Figure above.

Easy-to-Use Engineer Oriented Interfaces

Stepping through the analysis process can be very confusing to the non-specialist. However, with a next-generation CFD tool, the design engineer is stepped through the process by wizards that ensure all data such as materials, fluid characteristics, boundary conditions, and execution parameters are properly defined.

Multi Variate Parametric Modeling

This feature enables the design engineer to set up a series of experiments and submit multiple analyses for batch runs. The engineer can define a range of values and then modify the design (number of fins on the heat sink, dimensions of an outlet, etc.) or boundary conditions (pressure, flow volume, temperature, etc.) across a spectrum. The software automatically generates a series of analysis runs and submits them for execution. The results of these experiments can be compared and visualized, and then the best design can be chosen.

Laminar-Turbulent Transition Modeling

The modeling of a fluid flow is extremely complex while the flow transitions from laminar to turbulent or vice versa. In typical CFD software, this requires two different models and an understanding on the part of the engineer as to which one should be used. A model that senses this transition and automatically employs the correct modeling at the correct time in the transition eliminates the need for two different models and the intervention of the engineer.

Automated/Guaranteed Convergence

One of the most difficult tasks even for the CFD specialist is to ensure that the CFD analysis converges. Many factors can contribute to a CFD analysis non-convergence. Model complexity, meshing, boundary conditions, etc. can all cause older-generation software not to converge. The specialist must then experiment, sometimes using trial and error, to urge the software into convergence. With software such as FloEFD, convergence is virtually guaranteed, thus eliminating the need for the design engineer to tweak the model. These are just some examples of the technologies that enable complex CFD to be put into the hands of the design engineer as well as the specialist.

CFD Analysis for Complex Components to Systems

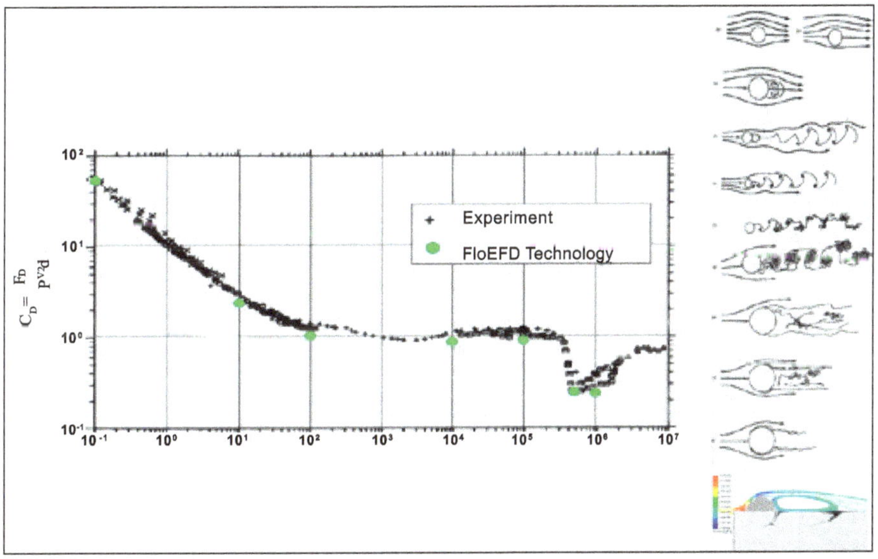

Analysis and experimental results are closely correlated using the laminar/turbulent transitional modeling.

When we think of CFD analysis, we might limit our thinking to the individual components of an aircraft or satellite, such as the electronics in a navigation computer for proper heat management, the fuel flow through an injector, or the capability of an oil cooling component. We might not imagine, given the size and complexity of a complete system such as fuel delivery, that we could analyze it with all the feet of piping, pumps, valves, etc. in an efficient manner. Using a traditional 3D CFD tool for this type of analysis would create a huge model requiring excessive computing power to analyze.

On the other hand, if we limited our analysis to the straight one dimensional (1D) flow of the fuel through the piping, we would compromise accuracy by not considering the effects of the complex components in the system. A good solution is to analyze the many feet of piping using a 1D CFD analysis software, while inserting the models for the more complex components using 3D CFD analysis.

Figure below shows an example of this type of system: an air-to-air refueling system with pipes and the refueling nozzle and how the combination of the 1D-3D analysis works.

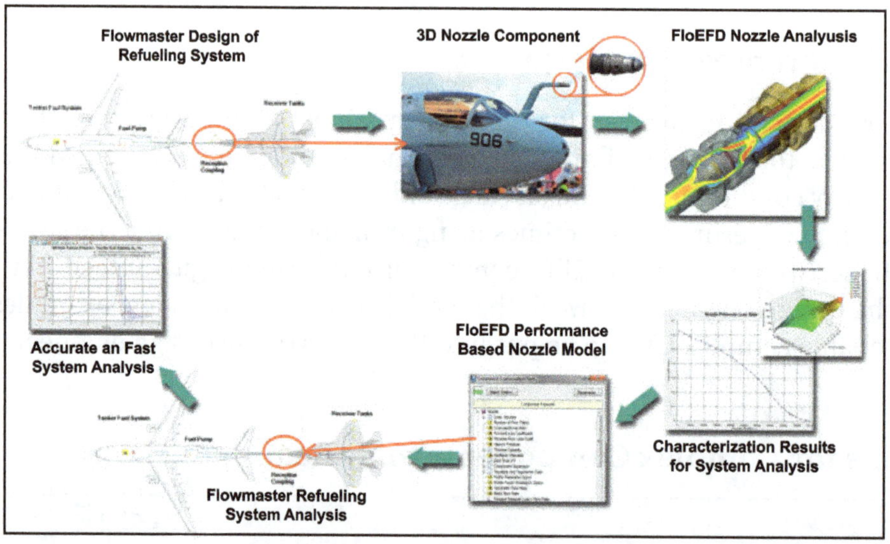

The combination of 1D and 3D CFD analysis can be used to analyze complex systems that contain both simple (piping) and complex (components) features.

The output from such an analysis could be a determination of the flow rates to refuel the jet or the "water hammer" effects if the aircraft were to suddenly disengage.

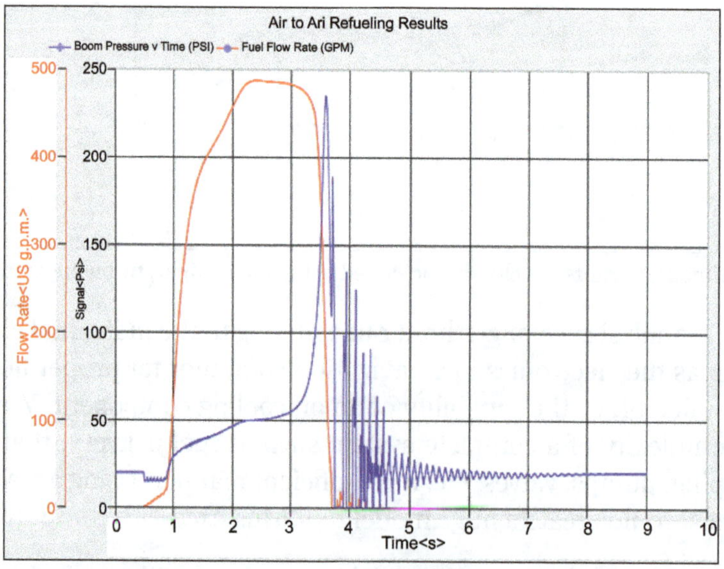

Analysis can calculate fuel flow rates (how long will it take to refuel the jet) and water hammer effects (will the system suffer damage if the aircraft suddenly disengage). This diagram shows the fuel flow rates in blue and the boom pressure in red.

Astrodynamics

Astrodynamics is the study of the motion in space of natural and artificial bodies upon which the major force is gravity. The resulting motion is the trajectory, whose selection is a key design parameter that influences the characteristics of the mission, including coverage or operational areas, ability to be accessed from Earth, launch vehicle selection, amount of time the spacecraft spends in sunlight, and its exposure to charged particle radiation. As a result, the trajectory has a profound effect on the requirements of most of the subsystems of a spacecraft. Experimentally, astrodynamics is founded on Kepler's laws of motion derived from observations of the motion of the planets with respect to the Earth. Kepler's laws are:

- The orbit of each planet is an ellipse with the Sun at one of the foci.
- The radius vector from the Sun to a planet sweeps out equal areas in equal times.
- The square of the period of a planet is proportional to the cube of the semimajor axis.

Kepler's second law is a result of the angular momentum being constant so that the speed must increase as the distance between bodies decreases. Mathematically, astrodynamics is based on Newton's laws of motion and his law for the gravitational force F_g between two masses of magnitudes m_s and m_e separated by distance r given by:

$$F_g = -Gm_s m_e / r^2$$

where $G = 6.67259 \times 10^{-11}$ m³/kg-s² is the universal gravitational constant. The relative motion of the two masses is given by:

$$r = a\left(1-e^2\right)/\left(1+e\cos f\right)$$
$$v = (k/m)\left[(2/r)-(1/a)\right]$$
$$E = -k/2a,\ e = \left(1+2H^2E/mk^2\right)^{1/2}$$
$$k \equiv Gm_1 m_2,\ m = m_s m_e/(m_s+m_e)$$

where a is the semimajor axis, e is the eccentricity, f is the true anomaly (the angular extent of one of the bodies from its point of closest approach), E is the total energy, H is the angular momentum, and m is known as the effective or equivalent mass. The equation for r is that of a conic section. For $e = 0$, Eq. above gives that $r = a =$ constant, i.e., a circular trajectory, where $E = -mk^2/2H^2$. For $0 < e < 1$, an elliptical trajectory results where $-mk^2/2H^2 < E < 0$. For $e = 1$, a parabolic trajectory results where $E = 0$, $a = \infty$, and $a(1 - e^2) \equiv p$, which is twice the distance from the point of closest approach to the focus. For $e > 1$, a hyperbolic trajectory results where $E > 0$ and $a < 0$. The period for the circular and elliptical trajectories and the perigee radius r_p (distance of closest approach to the focus) and apogee r_a (distance of farthest extent) for the elliptical trajectory follow from Equation above for $f = 0, \pi$ to be:

$$\tau = 2\pi\left(ma^3/k\right)^{1/2},\ r_p = a(1-e),\ r_a = a(1+e)$$

These equations give the characteristics of the orbit as a function of f, the true anomaly. To find the position as a function of time requires the introduction of a different transcendental equation for each type of trajectory. For the elliptical trajectory, it is necessary to solve Kepler's equation:

$$E = M + e \sin E$$

where,

$$r = a(1 - e\cos E), M \equiv nt, n \equiv \left[k / ma^3\right]^{1/2}$$

where E is the eccentric anomaly, M is the mean anomaly, n is the mean motion, and t is the time since perigee. The geometrical relationships between some of these parameters are given in Figure below. For a given time t, it is necessary to determine the mean motion n knowing the semimajor axis a, solve Equation above for the eccentric anomaly E, substitute E into Equation above to determine the radius r; and substitute r into Equation $r = a(1-e^2)/(1+e\cos f)...M \equiv nt, n \equiv \left[k / ma^3\right]^{1/2}$ to determine the true anomaly f; yielding r and f as functions of time. Since the distance to the circular orbit is constant, Equation above yield r = a, f = nt.

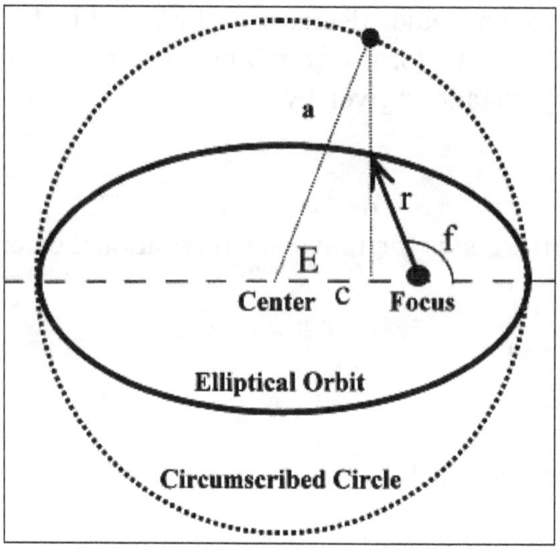

Elliptic orbit geometry for Kepler's Equation, where:

$$c = a - a(1-e) + r\,\mathrm{Cos} f$$
$$\mathrm{Cos} E = c / a = e + (r / a)\mathrm{Cos} f$$

Motion of a satellite around the Earth is defined in terms of Kepler's parameters, illustrated in Figures, where a is the semimajor axis, e the eccentricity, i the inclination, Ω the right ascension of the ascending node, ω the argument of perigee, and f the true anomaly. The right ascension of the ascending node Ω is measured from the equinox, which is the intersection of the ecliptic plane of the Earth and its equatorial plane, and is often identified as the first point of Aires ♈. If the Earth were spherical and homogeneous, the Kepler parameters, except for f, would be constant. However, the non-uniformity of the Earth's gravitational field causes Kepler's parameters to change as a function of time. The effects are generally categorized as short period, long period, and secular.

Secular effects are functions of powers of time, so they dominate. The primary effects due to the oblateness of the Earth are changes in Ω, ω, and M given by:

$$d\Omega / dt = 3J_1 nR^2 \cos i / \left(2a^2 \left(1-e^2\right)^2\right)$$

$$d\omega / dt = -3J_2 nR^2 \left(4-5\sin^2 i\right) / \left(4a^2 \left(1-e^2\right)^2\right)$$

$$dM / dt = n - J_1 nR^2 \left(1-(3/2)\sin^2 i\right) / \left(2a^2 \left(1-e^2\right)^{3/2}\right)$$

where for the Earth, $J_2 = -1.0827 \times 10^{-3}$ and R is its mean Equatorial radius. For very precise trajectory determination, it would be necessary to also take into account the perturbations of atmospheric drag, solar radiation pressure, and the gravitational force of the Sun and Moon. Atmospheric drag has the effect of circularizing an elliptical orbit and reducing its semimajor axis so that a low-altitude satellite will eventually reenter the Earth's atmosphere.

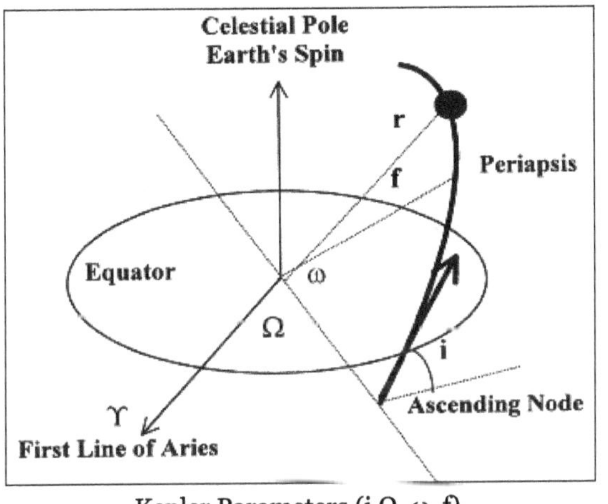

Kepler Parameters (i Ω, ω, f).

Orbit determination is the process of determining a set of initial conditions, such as the Kepler elements, that can be integrated in a software program or in a theoretical formulation to determine the most probable trajectory that best fit the observations. This is required to determine precisely where the spacecraft has been to correlate data that have been collected and to determine the future trajectory to predict observing times and whether changes in the trajectory will be required. Observations used include laser and radar range, radar range-rate, and one-way radio frequency Doppler collected by ground stations geographically dispersed. With the advent of the Global Positioning System (GPS) it possible to determine the trajectory in real time using an onboard GPS navigation system.

Propulsion

The propulsion element deals with the energy needed to move a vehicle through the air or in outer space. This energy is mostly provided by internal combustion engines, jet engines and turbomachinery, or rockets.

The operation of aerospace propulsion engines rests on the foundation of Newton's laws of motion.

The second of these laws explains that the change in momentum of the fluid passing through an engine is equal to the force acting on the fluid. The third law states that the force acting on the fluid exerts a reaction, an equal and opposite force, on the boundaries separating the fluid and the engine. Indeed, such engines are often referred to as reaction motors.

In general, aerospace propulsion engines may be thought of as idealized flow machines in which fluid within the machine has work and/or heat added to it prior to its exit from the machine as a jet, thereby producing thrust according to the reaction principle. The fluid may enter the machine from the surroundings or may be carried entirely within the machine prior to being processed. The former engines are usually called air breathing engines and the latter, rockets. Of primary interest is the magnitude of the thrust produced and the efficiency with which the heat and power is used in generating thrust.

The most efficient flow machine is the propeller, in which work is done on the fluid passing through it, but no heat is added. The source of power is external to the propeller itself and may be an internal combustion engine, a gas turbine, or even an electric motor. The propeller is speed-limited in the sense that for a given power input the thrust developed falls off with increasing speed.

Conversely, the turbojet, ramjet, scramjet, and pulsejet are flow machines in which heat is added to the airstream taken aboard, but no network is done on it. The heat is added by burning fuel in that airstream. The thrust they develop has a relatively weak dependence on flight speed making them capable of good performance at much higher speeds than are achievable with propellers. However, they are less efficient than propellers and their ability to travel at higher, even supersonic, speeds is thus gained at the cost of greater fuel consumption. The rocket, a variant of this case, takes in no ambient fluid but instead carries onboard all the fluid to which heat is added, again by combustion, and subsequently ejected as a jet. The thrust of the rocket is independent of the flight speed but this advantage is paid for by even lower efficiency than the other jet engines. On the other hand, because all the necessary propellant is carried onboard the rocket is able to operate in the vacuum of space.

Rather than add power alone or heat alone to the working fluid we show that by adding a combination of power and heat we can derive some of the special benefits of both the propeller and the turbojet. This is the turbofan engine which is the common jet powerplant for both commercial and military aircrafts.

Nuclear thermal propulsion (NTP) involves heating the onboard propellant by passing it through a nuclear fission reactor. Such rockets can provide high thrust with higher efficiency than a chemical rocket but pose radiation safety issues. Simple electrical resistance heating of a propellant is the basis of the resistojet while the arc jet uses an electric discharge to heat propellant. The highly efficient but low thrust ion rocket which generates thrust by electrostatic acceleration of ions and the magnetoplasmadynamic (MPD) rocket which uses electromagnetic acceleration to accelerate electric discharge-produced plasma. The ion and MPD rockets exhibit very high efficiency but are limited to relatively low thrust levels.

Aerodynamics

Aerodynamics is the study of how gases interact with moving bodies. Because the gas that we encounter most is air, aerodynamics is primarily concerned with the forces of drag and lift, which are caused by air passing over and around solid bodies. Engineers apply the principles of aerodynamics to the designs of many different things, including buildings, bridges and even soccer balls; however, of primary concern is the aerodynamics of aircraft and automobiles.

Aerodynamics comes into play in the study of flight and the science of building and operating an aircraft, which is called aeronautics. Aeronautical engineers use the fundamentals of aerodynamics to design aircraft that fly through the Earth's atmosphere.

Aerodynamic Drag

The most significant aerodynamic force that applies to nearly everything that moves through the air is drag. Drag is the force that opposes an aircraft's motion through the air, according to NASA. Drag is generated in the direction the air is moving when it encounters a solid object. In most cases, such as in automobiles and aircraft, drag is undesirable because it takes power to overcome it. There are, however, some cases when drag is beneficial, such as with parachutes, for example.

To describe the amount of drag on an object, we use a value called the drag coefficient (c_d). This number depends not only on the shape of the object but also on other factors, such as its speed and surface roughness, the density of the air and whether the flow is laminar (smooth) or turbulent. Forces that affect drag include the air pressure against the face of the object, the friction along the sides of the object and the relatively negative pressure, or suction, on the back of the object. For example, cd for a flat plate moving face-on through the air is about 1.3, a face-on cube is about 1, a sphere is about 0.5 and a teardrop shape is about 0.05. The drag coefficient for modern automobiles is 0.25 to 0.35, and for aircraft it is 0.01 to 0.03. Calculating cd can be complicated. For this reason, it is usually determined by computer simulations or wind tunnel experiments.

Aerodynamics of Aircraft

In order to overcome drag forces, an aircraft must generate thrust. This is accomplished with a motor-driven propeller or a jet engine. When the airplane is in level flight at a constant speed, the force of the thrust is just enough to counteract the aerodynamic drag.

Moving air can also generate forces in a different direction from the flow. The force that keeps an airplane from falling is called lift. Lift is generated by an aircraft wing. The path over a wing's curved top is longer than the path along the flat bottom of the wing. This causes the air to move faster over the top than it does along the bottom. With all other factors being equal, faster moving air has lower pressure than slower moving air, according to Bernoulli's principle, stated by Daniel Bernoulli, one of the most important pioneers in the field of fluid dynamics. This difference is what allows the slower moving air to push up against the bottom of the wing with greater force than the faster moving air is pushing down against the top of the wing. In level flight, this upward force is just enough to counteract the downward force caused by gravity.

Aerodynamic forces are also used to control an aircraft in flight. When the Wright brothers made their first flight in 1903, they needed a way to control their aircraft to climb, descend, bank and turn. They developed what is known as three-axis control for pitch, roll and yaw. Pitch (nose pointing up or down) is controlled by an elevator (the "flaps") on the back or trailing edge of the horizontal stabilizer in the tail section. Roll (tilting left or right) is controlled by ailerons (also flaps) on the trailing edges of the wings near the tips. Yaw (nose pointing left or right) is controlled by the rudder on the trailing edge of the vertical stabilizer in the tail section. These controls employ Newton's Third Law of Motion because they generate force by deflecting the airflow in the opposite direction of the desired movement. This force is also what allows aerobatic planes to fly upside down.

A pilot may also use flaps on the inboard section of the trailing edge of the wing during takeoff and landing. When in the downward position, flaps increase both lift and drag to allow the plane to fly slower without stalling. Some larger aircraft can also extend slats on the front or leading edges of the wings to increase lift at low speeds.

When the smooth airflow over a plane's wing is disrupted and this reduces the amount of lift, a stall can occur. "This is caused when the wing exceeds its critical angle of attack. This can occur at any airspeed, in any attitude, with any power setting." Typically, most stalls occur when an aircraft is moving too slowly with the nose at too high of an upward angle. The air no longer flows along the top surface but instead breaks away and forms turbulent swirls on top of the wing. This causes the plane to lose lift and start to fall, sometimes rather abruptly.

Another thing that can happen in an airplane is a spin. It is defined as "an aggravated stall that results in what is termed 'autorotation' wherein the airplane follows a downward corkscrew path." This usually occurs in a slow turn when the slower inside wing stalls, and the outside wing is still generating lift. "Especially at low altitude, successful spin recovery may be difficult if not impossible in many aircraft".

Control Engineering

Control system engineering or Control engineering is the branch of engineering which deals with the principles of control theory, to design a system which gives yields the desired behavior in a controlled manner. In aeronautics engineering, control means the tasks that need to be performed for achieving a desired result. The object to be controlled (for example a flight vehicle) is referred to as plant, while the process that exercises the control is called the controller. Both the plant and the controller are systems that are defined as self-contained sets of physical processes under study. A system working on known physical laws is said to be deterministic, whereas a system that runs on unknown (or partially known) physical laws is called nondeterministic, or stochastic. Every system has certain unwanted external input variables called disturbance inputs that cannot be modelled physically and thus, they are treated as stochastic disturbances. The disturbances are generally of two types:

- Process noise that can arise either due to unwanted external inputs or due to uncertainty in modelling the system.
- Measurement noise that results from uncertainty in measuring the output vector.

The presence of these external and internal imperfections renders all practical systems stochastic. State variables or state vector in vector form, are a set of scalar variables which show the condition or state of a system at a given time. State space is the vector space spanned by the state vector.

State of a system can be defined as a collection of the smallest number of variables necessary to completely specify the system's evolution in time in the absence of external inputs. Order of the system can be defined as the number of state variables required to represent a system, because it is equal to the net order of differential equations governing the system. The order of the state space (i.e., the size of the system) is unique, so any given system can be described by infinite alternative state-space representations.

Mathematical Equations Related to Aeronautics

A flight vehicle's state can be described by the position, r(t), velocity, v(t), angular velocity, ω(t), and orientation, ξ(t), relative to a frame of reference. Thus, the state vector of a flight vehicle's motion is:

$$x(t)=\{r(t),v(t),\omega(t),\xi(t)\}T$$

Depending upon the reference frame, x(t) can be transformed into any number of different state vectors. Control system consists of the plant and the controller. The controller manipulates the plant by sending a control input vector to the Plant. This vector is an output of the controller. Stating in physical terms, this output can take the form of either a force or a torque (or both) applied to a flight vehicle. Mostly, only electrical (or mechanical) signals are generated by the controller through wires (cables, hydraulic lines, etc.), that must be converted into physical inputs for the plant by a separate subsystem called an actuator.

But controllers generally require measurement of the output variables of the plant. If and when the measurement of a variable is involved, it is necessary to model the dynamics of the measurement process as a separate subsystem called a sensor. Talking actuators as there are measured scalar variables and scalar control inputs, respectively. Sensors and actuators can be modelled as part of either the plant or the controller.

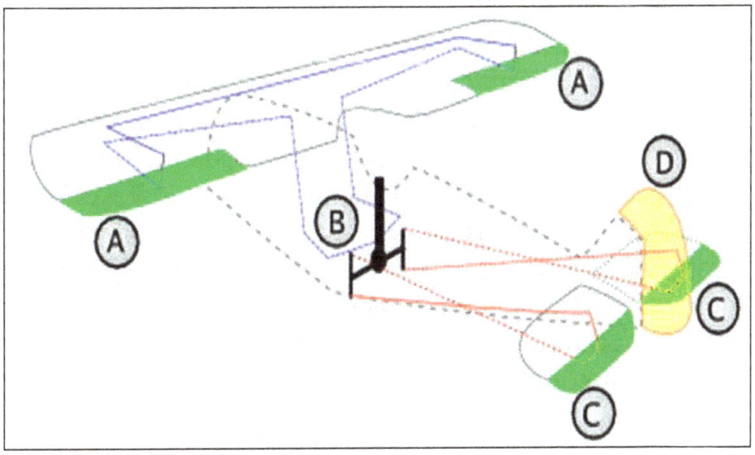

Motion Representation of an Aircraft.

To bring the plant to a desired state even in the presence of disturbances is the most common task of a control system. This can be achieved by either an open loop or a closed-loop control system. In an open-loop control system, controller has no knowledge of the actual state of the plant at a given time, and the control is exercised based on a model of the plant dynamics as well as an estimate of its state at a previous time instant, called the initial condition. Such an application of control can be successful in driving the plant to a desired state if and only if the plant model is exact and the external disturbances are absent, which is not possible generally. So, a closed-loop control system is the more practical alternative, in which the actual state of the plant is provided to the controller through a feedback loop, so that the control input u(t) can be adjusted appropriately. And in practice, the feedback consists of measurements of an output vector y(t) through which an estimate of the plant's state can be obtained by the controller. In case the feedback loop is removed, the control system becomes an open loop type.

Guidance and Control of Flight Vehicles

Flight vehicles are generally those vehicles that are capable of sustained motion through air or space and are broadly classified as aircraft, rockets and spacecraft. Aircraft flight is restricted to the atmosphere, and spacecraft flight to exo-atmospheric space, whereas rockets, which are usually operated in Earth's sensible atmosphere, are equally capable of operating both inside and outside it. Of these, aircraft achieve the lowest operating speeds due to the restriction imposed by aerodynamic drag (an atmospheric force opposing the motion and proportional to the square of the speed). Further, the requirement of deriving lift (a force normal to the flight direction necessary for balancing the weight) and thrust (the force along flight direction to counter drag) restricts aircraft flight to the smallest altitudes of all flight vehicles and prevents it from reaching the denser regions of the atmosphere. Contrary to this, spacecraft reach the highest speeds and altitudes because it generates centripetal acceleration entirely against gravity, while the operating speeds and altitudes of rockets (mostly employed for launching the spacecraft) fall somewhere between those of the aircraft and spacecraft.

Manipulation is required by all flight vehicles (i.e., adjustment or control) of position, velocity, and attitude (or orientation) for successful and efficient flight. A transport aircraft navigating between points A and B on the Earth's surface must follow a flight path that ensures the least possible fuel consumption in the presence of winds. Fighter aircrafts must be manoeuvred in a way that the normal acceleration is maximized while maintaining the total energy without exceeding the structural load limits. On the other hand, spacecraft launch rocket must achieve the necessary orbital velocity while maintaining a particular plane of flight. Missile rocket has to track a manoeuvring target such that an intercept is achieved before running out of propellant. A flight vehicle of atmospheric entry must land at a particular point with a specific terminal energy without exceeding the aerothermal load limits. Of all of the above cases, precise control of the vehicle's altitude is required at all times since the aerodynamic forces governing an atmospheric trajectory are very sensitive to the vehicle's orientation relative to the flight direction. Additionally, in some cases, attitude control alone is crucial for the mission's success. For example, a tumbling (or oscillating) satellite is useless as an observation or communications platform, even though it may be in the desired orbit. Similarly, a fighter (or bomber) aircraft requires a stable altitude for accurate weapons delivery.

Either a pilot or an automatic controller (auto-pilot) or both acting in tandem can achieve the flight vehicle control. Flying is termed as a process of controlling a flight vehicle. Manned vehicle is flown either manually by the pilot, or automatically by the autopilot that is programmed by the pilot to carry out a required task. Unmanned vehicles can be flown either remotely or by on board autopilots that receive occasional instructions from the ground. So, it is normal to have some form of human intervention in flying, and rarely do we have a fully automated (or autonomous) flight vehicle that not only performs the task that needs to be performed for a given mission, but also selects it. Therefore, the job of designing an automatic control system (or autopilot) is quite simple and consists of:

- Generating the required flight tasks for a given mission that are then stored in the autopilot memory as computer programs or specific data points.
- Putting in place a mechanism that closely performs the required (or reference) flight tasks at a given time, despite external disturbances and internal imperfections.

Radar Cross Section

The radar cross section (RCS) of a target is defined as the effective area intercepting an amount of incident power which, when scattered isotropically, produces a level of reflected power at the radar equal to that from the target. RCS calculations require broad and extensive technical knowledge, thus many scientists and scholars find the subject challenging and intellectually motivating. This is a very complex field that defies simple explanation, and any short treatment is only a very rough approximation.

The units of radar cross section are square meters; however, the radar cross section is NOT the same as the area of the target. Because of the wide range of amplitudes typically encountered on a target, RCS is frequently expressed in dBsm, or decibels relative to one square meter. The RCS is the projected area of a metal sphere that is large compared with the wavelength and that, if substituted for the object, would scatter identically the same power back to the radar. However, the RCS of all but the simplest scatterers fluctuates greatly with the orientation of the object, so the notion of an equivalent sphere is not very useful.

Different structures will exhibit different RCS dependence on frequency than a sphere. However, three frequency regimes are identifiable for most structures. In the Rayleigh region at low frequencies, target dimensions are much less than the radar wavelength. In this region RCS is proportional with the fourth power of the frequency. In the Resonance or Mie Region at medium frequencies, target dimensions and the radar wavelength are in the same order. The RCS oscillates in the resonance region. In the Optical Region of high frequencies, target dimensions are very large compared to the radar wavelength. In this region RCS is roughly the same size as the real area of target. The RCS behaves more simply in the high-frequency region. In this region, the RCS of a sphere is constant.

In general, codes based on the methods-of-moments (MOM) solution to the electrical field integral equation (EFIE) are used to calculate scattering in the Rayleigh and resonance regions. Codes based on physical optics (PO) and the physical theory of diffraction (PTD) are used in the optical or high-frequency region. The target's electrical size (which is proportional to frequency and inversely proportional to the radar wavelength) that determines the appropriate algorithm to calculate the scattering. When the target length is less than 5 to 10 wavelengths, the EFIE-MOM algorithm is used. Alternatively, if the target wavelength is above 5 to 10 wavelengths, the PO-PTD algorithm is used.

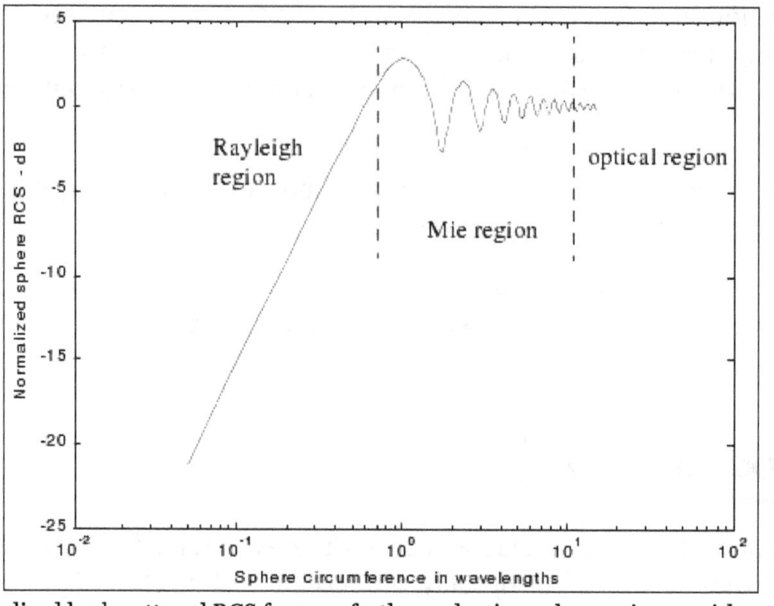

Normalized backscattered RCS for a perfectly conducting sphere using semi-log scale.

The RCS of a stealth aircraft is typically multiple orders of magnitude lower than a conventional plane and is often comparable to that of a small bird or large insect. "From the front, the F/A-22's signature is -40dBm2 (the size of a marble) while the F-35's is -30 dBm2 (the size of a golf ball). The F-35 is said to have a small area of vulnerability from the rear because engineers reduced cost by not designing a radar blocker for the engine exhaust." The F-35 stealthiness is a bit better than the B-2 bomber, which, in turn, was twice as good as that on the even older F-117. B-2 stealth bomber has a very small cross section. The RCS of a B-26 bomber exceeds 35 dBm2 (3100m2) from certain angles. In contrast, the RCS of the B-2 stealth bomber is widely reported to be about -40dBm2.

A conventional fighter aircraft such as an F-4 has an RCS of about six square meters (m2), and the much larger but low-observable B-2 bomber, which incorporates advanced stealth technologies into its design, by some accounts has an RCS of approximately 0.75 m2 [this is four orders of magnitude greater than the widely reported -40dBm2]. Some reports give the B-2 a head-on radar cross section no larger than a bird, 0.01 m2 or -20dBm2. A typical cruise missile with UAV-like characteristics has an RCS in the range of 1 m2; the Tomahawk ALCM, designed in the 1970s and utilizing the fairly simple low-observable technologies then available, has an RCS of less than 0.05 m2.

The impact of lowered observability can be dramatic because it reduces the maximum detection range from missile defenses, resulting in minimal time for intercept. The US airborne warning and control system (AWACS) radar system was designed to detect aircraft with an RCS of 7 m2 at a range of at least 370 km and typical non-stealthy cruise missiles at a range of at least 227 km; stealthy cruise missiles, however, could approach air defenses to within 108 km before being detected. If such missiles traveled at a speed of 805 km per hour (500 miles per hour), air defenses would have only eight minutes to engage and destroy the stealthy missile and 17 minutes for the non-stealthy missile. Furthermore, a low-observable LACM can be difficult to engage and destroy, even if detected. Cruise missiles with an RCS of 0.1 m2 or smaller are difficult for surface-to-air missile (SAM) fire-control radars to track. Consequently, even if a SAM battery detects the missile, it may not acquire a sufficient lock on the target to complete the intercept.

Radar scattering from any realistic target is a function of the body's material properties as well as its geometry. Once the specular reflections have been eliminated by radar absorbing materials, only non-specular or diffractive sources are left. Non-specular scatterers are edges, creeping waves, and traveling waves. They often dominate backscattering patterns of realistic targets in the aspect ranges of most interest. The traveling wave is a high frequency phenomenon. Surface traveling waves are launched for horizontal polarization and grazing angles of incidence on targets with long smooth surfaces. There is little attenuation from the flat smooth surface, so the wave builds up as it travels along the target. Upon reaching a surface discontinuity, for example an edge, the traveling wave is scattered and part of it propagates back toward the radar. The sum of the traveling waves propagating from the far end of the target toward the near end is the dominant source to the target radar cross section.

The radar cross section (RCS) of a target not only depends on the physical shape and its composite materials, but also on its subcomponents such as antennas and other sensors. These components on the platforms may be designed to meet low RCS requirements as well as their sensor system requirements. In some cases, the onboard sensors can be the predominant factor in determining a platform's total RCS. A typical example is a reciprocal high gain antenna on a low RCS platform. If the antenna beam is pointed toward the radar and the radar frequency is in the antenna operating band, the antenna scattering can be significant.

The traditional measure of an object's scattering behavior is the RCS pattern which plots the scattered field magnitude as a function of aspect angle for a particular frequency and polarization. Although suitable to calculate the power received by a radar operating with those particular parameters, the RCS pattern is an incomplete descriptor of the object's scattering behavior. While the RCS pattern indicates the effect of the scattering mechanism, it does not reveal the physical processes which cause the observed effect. In contrast, imaging techniques, which exploit frequency and angle diversity to spatially resolve the reflectivity distribution of complex objects, allow the association of physical features with scattering mechanisms. These processes, therefore, indicate the causal components of the overall signature level observed in RCS patterns.

Aeroelasticity

Structural dynamics is a wide scientific domain covering a very large range of applications such as structure vibrations, crash and impact, structure damping, vibroacoustics and so on. Aeroelasticity is strongly linked to the subdomain of the structural dynamics dedicated to the low frequencies. It can be briefly defined as the study of the dynamic behavior of a structure (aircraft, turbomachine, helicopter rotor, etc.) in an aerodynamic flow, or, as "the study of the mutual interaction that takes place within the triangle of the inertial, elastic and aerodynamic forces acting on structural members exposed to an airstream, and the influence of this study on design".

Though Dutch windmill manufacturers empirically solved blade aeroelastic problems four centuries ago by moving the front spar from the mid-chord to the quarter-chord position, aeroelasticity is a rather recent discipline that arose with the first flight attempts.

After the success of the Wright flight, it was mainly biplanes that were developed in order to ensure a high torsional stiffness of the wings, but some aircraft as the Handley Page 0/400 bomber and the DH-9 still experienced tail flutter instabilities. One of the earliest documented cases mentioning failure resulting from the static interaction between the air flow and the wing deformation is that of the monoplane Fokker D-8, developed for its superior performance compared to biplanes during World War 1.

The aeroelasticity scientific domain has grown up alongside the aviation boom, especially during and after World War 2, with the increasing flight speed of the aircraft. It now concerns all aeronautical applications (aircraft, turbomachines, helicopters, launchers, missiles) as well as others such as wind turbines, bridges and buildings.

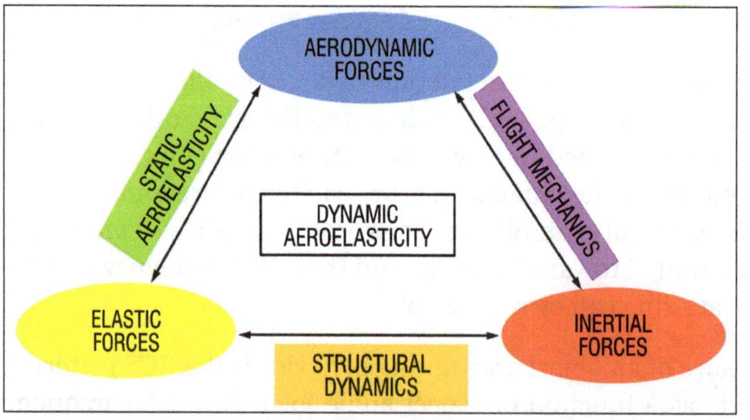

Collar's representation of forces occurring in aeroelastic phenomena.

Aeroelasticity used to be taken into account at a late stage of an airplane design, to check its stability. But today aeronautics has to drastically reduce its impact on the environment. One way to reach this goal is to decrease fuel consumption. For this purpose, a considerable effort is made to decrease the weight of the structures, which thereby tend to become more flexible. Furthermore, numerous new concepts of aircraft such as low sweep angle and high aspect ratio wing, laminar wing, highly flexible wing and strut braced wing, to mention just a few, are being developed to improve aerodynamic performance. But these new concepts tend toward lighter and more flexible wings. Therefore today aeroelasticity has an increasing influence on the design and has to be taken into account at earlier or even preliminary stages, not only to check the stability of the airplane design but also to compute its actual aerodynamic performance, the control surface effectiveness, and so prevent its reversal, and of course to verify both the static (static divergence) and dynamic (flutter) stabilities. Multi-disciplinary design and optimization methods are currently being developed to simultaneously determine the aerodynamic shape and the structural stiffness taking into account interactions between the aerodynamic flows and the structure.

Avionics

Avionics is defined the development and production of electronic instruments for use in aviation and astronautics. The term also refers to the instruments themselves. Such instruments consist of a wide variety of control, performance, communications, and radio navigation devices and systems. Avionics includes all instruments, sensors, and electronic equipment and the electrical systems that link them to each other and to aerospace vehicle-control systems. It encompasses the functional equipment for guidance, navigation, and communications. A modern airline transport can contain more than 1,000 sensors and "black boxes." The latter are metal or plastic housings in which electronic and electrical components are grouped to perform specific functions. (They differ from flight recorders, also dubbed black boxes, which record the performance and condition of an aircraft in flight.) For advanced military aircraft, avionics represents as much as 35 percent of the total cost; when radar and other electronic and electro-optic system adjuncts are included, the value can exceed 50 percent. For some spacecraft, the equivalent equipment can reach 70 percent of the cost.

Aeroacoustics

The generation of sound by a turbulent flow is the most common physical effect associated with the field of aeroacoustics. The prefix aero means air, however, the field of aeroacoustics is not restricted to flow-induced noise in air. Aeroacoustics is concerned with the general interaction between a background flow and an acoustics field. For example, if you were studying the reflections of sound off a shear layer or how the flow in a muffler affects transmission loss, you would be studying aeroacoustics.

Aeroacoustics models can include effects that alter an acoustic field in the presence of a flow, such as turbulence, local variations in material properties, convection, viscous damping, and much more. Numerically solving aeroacoustic problems falls under the field of computational aeroacoustics (CAA).

Flow-Induced Noise

Acoustic noise generated by a flow can be created through different mechanics, but is ultimately due to fluctuations in the flow. These fluctuations will give rise to distributed acoustic sources throughout the flow. Noise is created by local stress fluctuations in the flow (Reynolds stresses, viscous stress effects, and nonisentropic effects all act as quadrupole sources); pressure fluctuations at walls (e.g., dipole sources at solid boundaries); mass and heat fluctuations (e.g., distributed monopole sources); and external fluctuating force fields.

Acoustic waves are only a subset of waves that are created by the flow; vorticity and thermal instability waves are also produced. These particular waves are only convected by the flow, whereas acoustic waves also propagate relative to the flow, at the local speed of sound. Compared to other fluctuations, the energy associated with acoustic waves is typically many orders of magnitude smaller. This means that a direct simulation of the flow-induced noise is very challenging and requires extremely accurate numerical schemes. Another difficulty is that the acoustic sources occur at the length scale of the turbulence in the flow, which then needs to be resolved.

Acoustic Analogies

Acoustic analogy equations are formulations that separate out acoustic fluctuations from flow fluctuations.

The first form of these equations derived from the seminal work of J. Lighthill in the 1950s. His acoustic analogy equation is obtained by rearranging the Navier-Stokes equations, which result in a convected scalar wave equation for the acoustic pressure and a right-hand side that contains all of the flow source terms.

Analogy equations can also be constructed by rearranging the flow equations in other ways, leading to linearized Euler equations or linearized Navier-Stokes equations with right-hand side flow sources. The linearized models have great applicability in many aeroacoustics situations where the flow-induced noise is not important. This is found in fluid-structure interaction (FSI) problems in the frequency domain, when modeling how a background flow affects the acoustics (e.g., reflection and refraction), detailed muffler models, and much more.

Linearized Model of Aeroacoustics

Ideally, aeroacoustic simulations would involve solving the fully compressible Navier-Stokes equations in the time domain. Acoustic pressure waves would then form a subset of the fluid solution. However, this approach is often impractical for real-world applications, due to the computational time and memory resources required. Instead, like solving for many practical engineering problems, a decoupled two-step multiphysics approach is used:

- Solve for the fluid flow.
- Solve for the acoustic perturbations on top of the flow.

Several models exist that include different levels of detail and different aeroacoustic effects. These models are derived by linearizing the dependent variables around a steady-state background flow and making physical assumptions. The dependent variables now represent perturbations, such as acoustic and instability waves, on top of the flow. The classical scalar wave equation, otherwise known as the Helmholtz equation in the frequency domain, is an example of a linearized model with no explicit losses and an isentropic assumption.

Types of Linearized Model

Scalar Wave Equation

In its general form, the scalar wave equation includes the dependency of the speed of sound and density on the background flow (as they can both be spatially varying). It is a good approximation when convective effects are not important (i.e., when the Mach number is less than 0.1) and thermal and viscous losses are also negligible.

Thermoacoustics

Linearized thermoacoustics models involve the dependency of material parameters on the background flow, as well as thermal and viscous losses. They are sufficient models when the Mach number is less than 0.1. An application example is modeling FSI in the frequency domain.

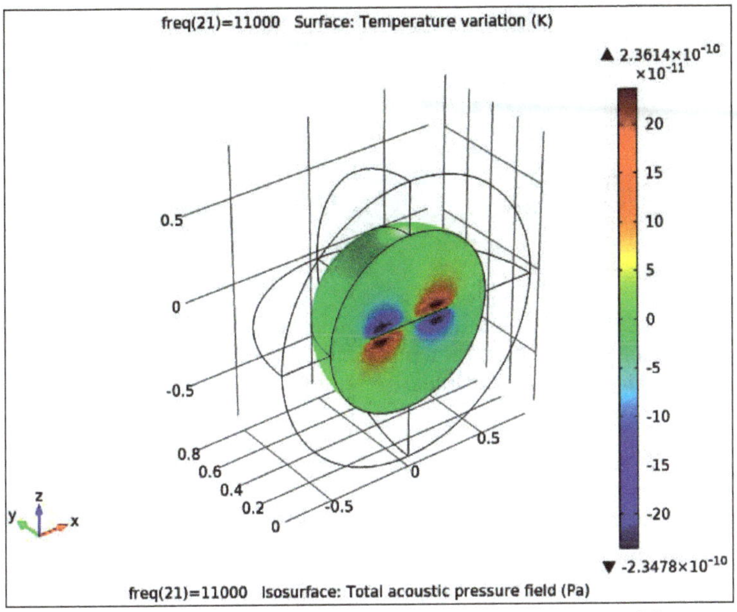

Model of a vibrating micromirror. The plot shows the acoustic variations of the temperature field.

Linearized Potential Flow

Linearized potential flow models involve the interaction between a background potential flow and the acoustic field. Here, no losses are modeled except for possible impedance conditions at walls. Linearized potential flow can be used for modeling sound propagation from jet engines, for example.

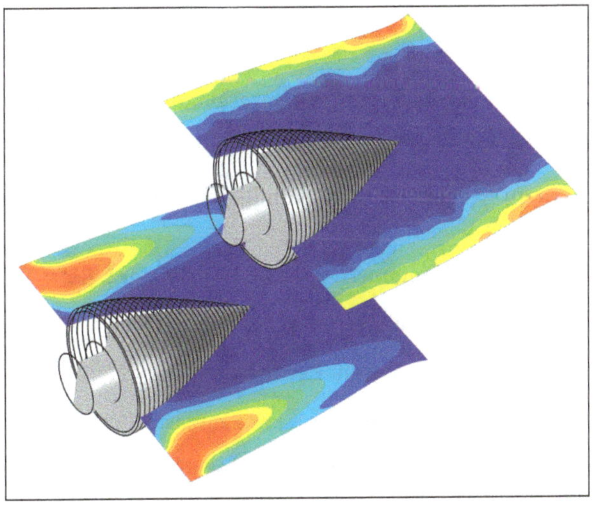

Acoustic pressure distribution on a lined duct wall of a jet engine.

Linearized Euler

Linearized Euler models can be used to compute acoustic variations in density, velocity, and pressure in the presence of a stationary background mean flow that is well approximated by an ideal gas flow. In its general form, this equation supports both acoustic and non-acoustic waves. The latter are so-called instability waves (vorticity and entropy waves).

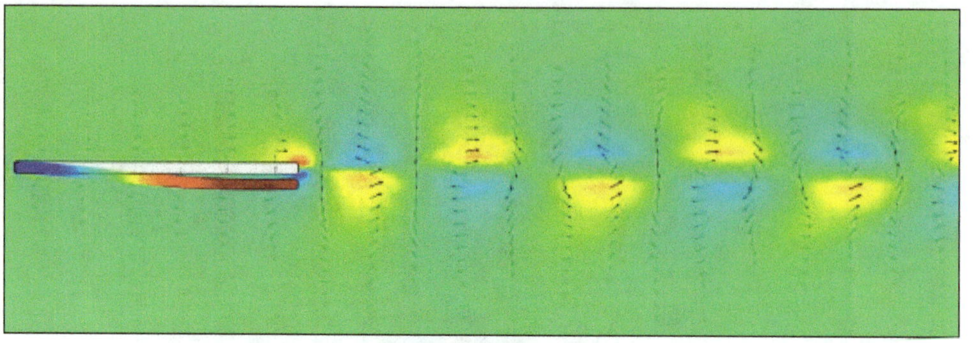

Visualizing the vibrations of a plate in a 2D viscous parallel plate flow.
Modeled with the linearized Navier-Stokes equations.

Linearized Navier-Stokes

Linearized Navier-Stokes equations are used to compute the acoustic variations in pressure, velocity, and temperature in the presence of any stationary isothermal or nonisothermal background mean flow. These equations include viscous losses, thermal conduction, and heat generated by viscous dissipation, if relevant.

Acoustic Perturbation Equations

There are many variations and modifications of the aforementioned linearized models for modeling aeroacoustics problems. One such example is the acoustic perturbation equations. Here, non-acoustic propagating modes have been filtered out by modifying the governing equation, resulting in a pure acoustic problem.

References

- Aerospace-engineering, engineering-branch: indiaeducation.net, Retrieved 21, July 2020
- Fluid-dynamics-47446: livescience.com, Retrieved 13, March 2020
- Engineering-Fluid-Mechanics-Mechanical-Engineering-Handbook-Kreith-Berger-26605708: academia.edu, Retrieved 25, June 2020
- Cfd-in-the-aerospace-and-aeronautics-industries-2168: longdom.org, Retrieved 09, April 2020
- Aerospace-propulsion: sciencedirect.com, Retrieved 29, August 2020
- What-is-aerodynamics-47930: livescience.com, Retrieved 03, May 2020
- Control-Engineering-in-Aeronautic-275333574: researchgate.net, Retrieved 17, February 2020
- Research-on-aeroelasticity-phenomenon-in-aeronautical-engineering, aerodynamics: intechopen.com, Retrieved 05, June 2020
- Secondary-and-tertiary-aerospace-systems, aerospace-industry- technology-753785: britannica.com, Retrieved 24, March 2020

2

Types of Aircraft

The types of aircraft can be broadly classified into lighter than air systems, heavier than air systems, and gliders and sailplanes. Airships and balloons are a few types of lighter than air systems while helicopters and fixed wing aircrafts fall under heavier than air systems. This chapter discusses these types of aircraft in detail.

There are several ways to classify aircraft. Below, we describe classifications by design, propulsion and usage.

By Design

A first division by design among aircraft is between lighter-than-air, aerostat, and heavier-than-air aircraft, aerodyne. Examples of lighter-than-air aircraft include non-steerable balloons, such as hot air balloons and gas balloons, and steerable airships (sometimes called dirigible balloons) such as blimps (that have non-rigid construction) and rigid airships that have an internal frame. The most successful type of rigid airship was the Zeppelin. Several accidents, such as the Hindenburg fire at Lakehurst, NJ, in 1937 led to the demise of large rigid airships.

A collection of NASA test aircraft.

In heavier-than-air aircraft, there are two ways to produce lift: Aerodynamic lift and engine lift. In the case of aerodynamic lift, the aircraft is kept in the air by wings or rotors. With engine lift, the aircraft defeats gravity by use of vertical thrust. Examples of engine lift aircraft are rockets, and VTOL aircraft such as the Hawker-Siddeley Harrier.

Among aerodynamically lifted aircraft, most fall in the category of fixed-wing aircraft, where horizontal airfoils produce lift, by profiting from airflow patterns determined by Bernoulli's equation and, to some extent, the Coanda effect. The forerunner of these type of aircraft is the kite. Kites depend upon the tension between the cord which anchors it to the ground and the force of the wind currents. Much aerodynamic work was done with kites until test aircraft, wind tunnels and now computer modelling programs became available.

In a "conventional" configuration, the lift surfaces are placed in front of a control surface or tailplane. The other configuration is the canard where small horizontal control surfaces are placed forward of the wings, near the nose of the aircraft. Canards are becoming more common as supersonic aerodynamics grows more mature and because the forward surface contributes lift during straight-and-level flight.

The number of lift surfaces varied in the pre-1950 period, as biplanes (two wings) and triplanes (three wings) were numerous in the early days of aviation. Subsequently most aircraft are monoplanes. This is principally an improvement in structures and not aerodynamics.Other possibilities include the delta wing, where lift and horizontal control surfaces are often combined, and the flying wing, where there is no separate vertical control surface (e.g., the B-2 Spirit). A variable-geometry wing (or "swing-wing") has also been employed in a few examples of combat aircraft, such as the F-111, Panavia Tornado, F-14 Tomcat and B-1 Lancer, among others.

The lifting body configuration is where the body itself produce lift. So far, the only significant practical application of the lifting body is in the Space Shuttle, but many aircraft generate lift from nothing other than wings alone.A second category of aerodynamically lifted aircraft are the rotary-wing aircraft. Here, the lift is provided by rotating aerofoils or rotors. The best-known examples are the helicopter, the autogyro and the tiltrotor aircraft (such as the V-22 Osprey). Some craft have reaction-powered rotors with gas jets at the tips but most have one or more lift rotors powered from engine-driven shafts.

A further category might encompass the wing-in-ground-effect types, for example the Russian ekranoplan also nicknamed the "Caspian Sea Monster" and hovercraft; most of the latter employing a skirt and achieving limited ground or water clearance to reduce friction and achieve speeds above those achieved by boats of similar weight.

A recent innovation is a completely new class of aircraft, the fan wing. This uses a fixed wing with a forced airflow produced by cylindrical fans mounted above. And finally the flapping-wing ornithopter is a category of its own. These designs may have potential, but currently have no major practical applications.

By Propulsion

Some types of aircraft, such as the balloon or glider, do not have any propulsion. Balloons drift with the wind, though normally the pilot can control the altitude either by heating the air or by releasing ballast, giving some directional control (since the wind direction changes with altitude). For gliders, takeoff takes place from a high location, or the aircraft is pulled into the air by a ground-based winch or vehicle, or towed aloft by a powered "tug" aircraft. Airships combine a balloon's buoyancy with some kind of propulsion, usually propeller driven.

Gossamer Albatross, a human-powered aircraft.

Until World War II, the internal combustion piston engine was virtually the only type of propulsion used for powered aircraft. The piston engine is still used in the majority of aircraft produced, since it is efficient at the lower altitudes used by small aircraft, but the radial engine (with the cylinders arranged in a circle around the crankshaft) has largely given way to the horizontally-opposed engine (with the cylinders lined up on two sides of the crankshaft). Water cooled V engines, as used in automobiles, were common in high speed aircraft, until they were replaced by jet and turbine power. Piston engines typically operate using avgas or regular gasoline, though some new ones are being designed to operate on diesel or jet fuel.

Piston engines normally become less efficient above 7,000-8,000 ft (2100-2400 m) above sea level because there is less oxygen available for combustion; to solve that problem, some piston engines have mechanically powered compressors (blowers) or turbine-powered turbochargers or turbonormalizers that compress the air before feeding it into the engine; these piston engines can often operate efficiently at 20,000 ft (6100 m) above sea level or higher, altitudes that require the use of supplemental oxygen or cabin pressurization. During the forties and especially following the 1973 energy crisis, development work was done on propellers with swept tips or even scimitar-shaped blades for use in high-speed commercial and military transports.

Pressurised aircraft, however, are more likely to use the turbine engine, since it is naturally efficient at higher altitudes and can operate above 40,000 ft. Helicopters also typically use turbine engines. In addition to turbine engines like the turboprop and turbojet, other types of high-altitude, high-performance engines have included the ramjet and the pulse jet. Rocket aircraft have occasionally been experimented with. They are restricted to rather specialised niches, such as spaceflight, where no oxygen is available for combustion (rockets carry their own oxygen).

By Usage

The major distinction in aircraft usage is between military aviation, which includes all uses of aircraft for military purposes (such as combat, patrolling, search and rescue, reconnaissance, transport, and training), and civil aviation, which includes all uses of aircraft for non-military purposes.

Military Aircraft

Combat aircraft like fighters or bombers represent only a minority of the category. Many civil aircraft have been produced in separate models for military use, such as the civil Douglas DC-3 airliner, which became the military C-47/C-53/R4D transport in the U.S. military and the "Dakota" in the UK and the Commonwealth. Even the small fabric-covered two-seater Piper J3 Cub had a military version, the L-4 liaison, observation and trainer aircraft. In the past, gliders and balloons have also been used as military aircraft; for example, balloons were used for observation during the American Civil War and World War I, and cargo gliders were used during World War II to land troops.

Combat aircraft themselves, though used a handful of times for reconnaissance and surveillance during the Italo-Turkish War, did not come into widespread use until the Balkan War when first air-dropped bomb was invented and widely used by Bulgarian Air Force against Turkey.

On January 24 1913 the first naval co-operation mission took place by Greek Air Force planes above the Dardanelles. During World War I many types of aircraft were adapted for attacking the ground or enemy vehicles/ships/guns/aircraft, and the first aircraft designed as bombers were born. In order to prevent the enemy from bombing, fighter aircraft were developed to intercept and shoot down enemy aircraft. Tankers were developed after World War II to refuel other aircraft in mid-air, thus increasing their operational range. By the time of the Vietnam War, helicopters had come into widespread military use, especially for transporting, supplying, and supporting ground troops.

Civil Aviation

Civil aviation includes both scheduled airline flights and general aviation, a catch-all covering other kinds of private and commercial use. The vast majority of flights flown around the world each day belong to the general aviation category, ranging from recreational balloon flying to civilian flight training to business trips to firefighting to medevac flights to cargo transportation on freight aircraft.

Interior of Qatar Airways flight.

Within general aviation, the major distinction is between private flights (where the pilot is not paid for time or expenses) and commercial flights (where the pilot is paid by a customer or employer). Private pilots use aircraft primarily for personal travel, business travel, or recreation. Usually these private pilots own their own aircraft and take out loans from banks or specialized lenders to purchase them. Commercial general aviation pilots use aircraft for a wide range of tasks, such as flight training, pipeline surveying, passenger and freight transport, policing, crop dusting, and medical transport (medevac). Piston-powered propeller aircraft (single-engine or twin-engine) are especially common for both private and commercial general aviation, but even private pilots occasionally own and operate helicopters. Business jets are typically flown by commercial pilots, although there is a new generation of small jets arriving soon for private pilots.

Bell 206B JetRanger III helicopter.

Lighter-than-Air Systems

Lighter-than-Air (LTA) systems belong to a class of aerospace platforms that use buoyancy of a lighter-than-air gas to generate bulk of the lift force to overcome gravity; hence they are able to operate even in the absence of any relative motion between the platform and ambient air. Though LTA systems suffer from operational limitations, they do have some unique capabilities, which can make them a platform of choice for certain applications.

The most commonly known LTA platforms are hot-air or gas-filled balloons; but since these do not have any worthwhile mechanism to steer them, they are of limited use for most applications. Aerostats and Airships are more capable LTA platforms; Aerostats are aerodynamically shaped tethered balloons mounted with fins for stability. Airships are untethered; and since they have on-board propulsion and control system, they can handle changes in ambient wind much better, and hence can relocate and fly around.

Both Aerostats and Airships can act as high-altitude platforms for several scientific and commercial applications. They have been used all over the globe (and India) for long endurance aerial observation, imaging and communications, as well as an aerial banner for advertisement and product promotion, for which they are uniquely suitable. In fact, LTA systems are most suited to meet the requirement of a slow moving or nearly stationary aerial platform with very long endurance and low vibration levels, to be deployed in mostly in fair weather and calm ambient conditions. They have excellent green credentials, since their fuel consumption and power requirements are a small fraction of that of any other aerial vehicle for these applications.

Applications of LTA Systems

Aerostats have been actively applied to various problems. These include aerial imaging, remote sensing, radar visual and infrared monitoring of international borders, airspace and movement on the ground , traffic monitoring and control, synthetic aperture astronomical telescopes, relaying electromagnetic signals, and generation of wind power from the jet stream, and collection of solar power from above the cloud layer. Researchers have proposed power generation from aerodynamic kites, but many of the application lessons may be applied and scaled up to use aerostats which can guarantee liftoff from the ground and endurance regardless of ground-level or upper-level winds.

Relevance to Retail Power Beaming

When the issues in power beaming are examined, two things become clear. The first is that diffraction-limited antenna size becomes unacceptably large for rural receivers, at frequencies below 100GHz. Where low frequencies are used, antennae are sized to receive no more than the central main lobe of the beam, thus losing roughly 16% of the transmitted power just through spillage. The second point is that above 10GHz, horizontal power beaming at low altitudes incurs extreme losses, and is impractical beyond 1 or 2 kilometers.

For adoption of the 200-225GHz band for beaming vertically through the atmosphere, and perhaps a lower waveband such as 100GHz for horizontal transmission over short distances. In their proposed architecture for rural power beaming in India, local wired grids would be used within a village and its immediate surroundings. Lower-frequency wireless beaming would be used over short distances where the antenna size and transmission loss are not prohibitive, if justified by a tradeoff with the immediate costs and the opportunity costs of installing a wired grid for that purpose.

They argue that a wired grid will follow, dictated by the economics of power cost, once the local economy picks up using wireless power for the startup phase. To deliver power over distances of several tens of kilometers, they propose high-altitude (stratospheric) platforms, accompanied by lower-altitude aerostats. These enable the power to be beamed vertically up and down, minimizing distance transmitted through the moist, dense, lower troposphere, and then horizontal line-of-sight beaming through the clear, dry, low-density upper atmosphere using 220GHz beaming.

Their interest stems from considering the retail power beaming end of a comprehensive Space Power Grid architecture, enabling renewable power plants around the world to exchange power through and between Space satellites. They have considered several of the difficult issues in millimeter wave power beaming in earlier work. While there are obviously difficult obstacles in making millimeter wave power generators and receivers for such high-power applications, and in transmitting such power efficiency, the conversation in power beaming and Space Solar Power has clearly shifted to the regime of millimeter waves. As pointed out above, LTA platforms at several levels, provide key enablers for power beaming.

Airships

In airship design three different construction methods need to be differentiated. The differences can be seen mainly on airship structures and their lifting gas management systems. These aircraft are being separated into blimps, semi-rigid and rigid airships.

Types of Airships

Blimp

Blimps do not have an internal support structure as the other airship types. Stability is being achieved by a slight over pressure of the lifting gas within the hull of about 0.05 to 0.1 bar. The gondola, engines and other subsystems are being fixed at the hull through catenary wires. Those non-rigid airships are often being called blimps. These aircraft have the big disadvantage that they are subject to size constraints due to stability problems.

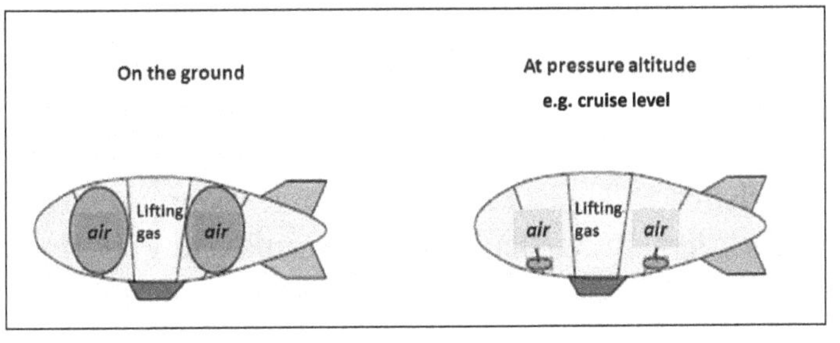

Semi-rigid Airship

Semi-rigid airships combine the construction methods of blimps and rigid airships. This type consists of a lifting gas hull, which is fixed at an inside or outside keel. Exactly as blimps, semi-rigid airships are being stabilized through the ballonet-technique. A huge advantage of this particular construction method is the solid fixation of gondola, engines and control surfaces. As today, this design represents the most common construction method for new airships. The most prominent exponents are the Zeppelin NT and the never built Cargolifter CL 160.

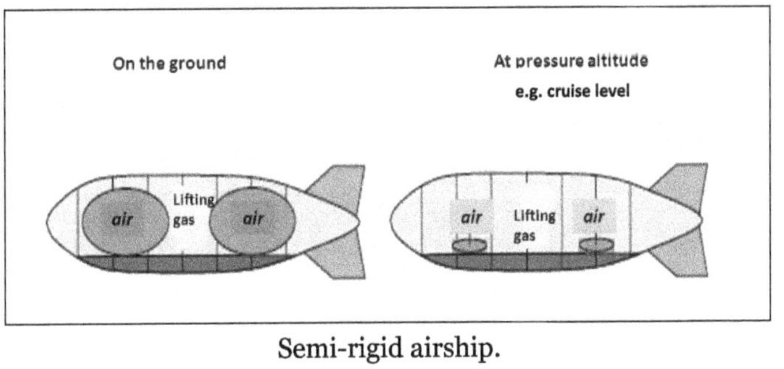

Semi-rigid airship.

Rigid Airship

This specific type of airship has a rigid structure, which dictates the external shape of the aircraft. This frame contains gas cells, which expand with increasing altitude and are enclosed by an outer hull. The space in between is filled with air. Due to a very high rigidity of that structure, gondolas, control surfaces, engines and other systems can be attached quasi arbitrarily on that framework. The widely known Graf Zeppelin or Hindenburg aircraft were rigid airships. Besides all of those conventional airship designs a few hybrid solutions have been identified since the 1970s.

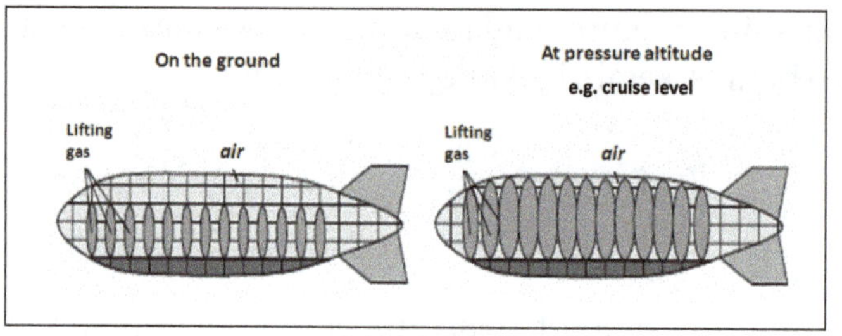

Rigid-airship.

Hybrid Airship

An aircraft concept combining the features of a helicopter or a fixed wing airplane with those of an airship is being called a hybrid airship. Experts define a concept as a hybrid airship when more than 40% of the weight is being carried by dynamic lift. But such an airship does not represent a proven alternative to today's air transport systems. Airship manufacturers wish for new application scenarios for those concepts in the future. Due to higher cruising speeds, this flying body has quite stable flight characteristics and better maneuverability compared to regular airships. Therefore its dependability on weather, in particular heavy winds and gusts is lower. Furthermore the usual ground infrastructure as used for conventional airships is not needed anymore. This is due to Heavier-Than-Air (HTA) features of that aircraft, which has the ability to take off and land similar to an airplane. Santos Dumont was the first, who had the idea of such a hybrid airship concept in 1905. Figure below shows the terminology of the flight system aircraft.

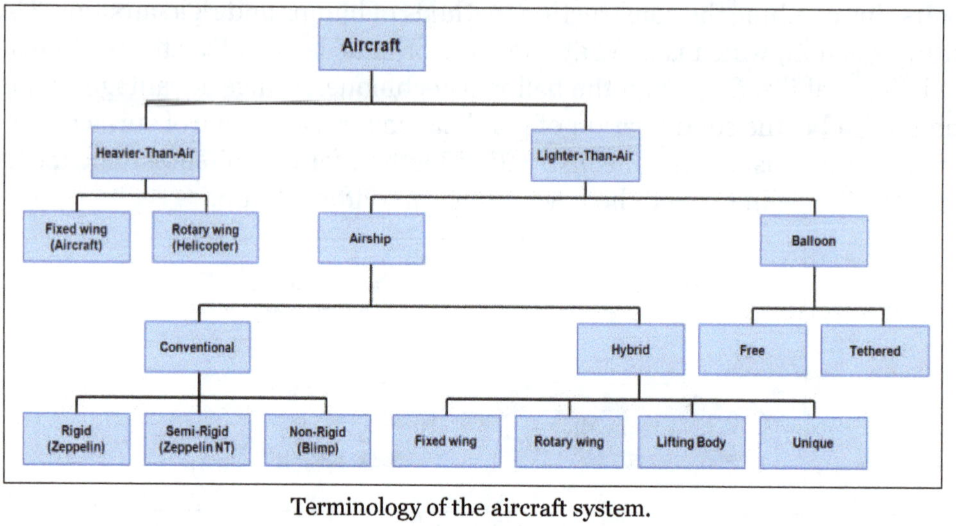

Terminology of the aircraft system.

Chronology and Former Application

Airship Technology

Airships and the Lighter-Than-Air technology in general have a long history behind them. The man's dream of being able to fly was realized by the Montgolfière brother in 1783. The idea not to fly in a balloon, but to lift off in an airship came from General J. Meusnier in the year 1795. Figure shows his concept, which already had a great similarity to airships, designed 100 years later. Unfortunately his construction has never been realized. The airship development has been pushed quite rapidly since the 19th century. Size, payload and speed rose constantly during the years. The size of the spheres in figure below describes the respective payload capacity.

Table: Speed and payload development of airships.

S.No	Type	Year	Speed (km/h)	Payload (kg)	Built
1	LZ 1	1900	28	1400	Yes
2	LZ 6	1910	56	4400	Yes
3	LZ 36	1915	85	11000	Yes
4	LZ 120	1920	127	96000	Yes
5	LZ 129	1936	137	800	Yes
6	WDl 1B	1988	105	800	Yes
7	American Blimp A-150	1997	95	628	Yes
8	Skyship 600	1998	105	800	Yes
9	Spirit of Goodyear GZ 20A	2000	80	650	Yes
10	Zeppelin NT	2000	125	1900	
11	Cargolifter CL 160	2005	125	160000	No
12	Aeros 40 D	2006	82	980	
13	Skycat 20	2008	157	20000	No
14	Aeroscraft ML16	2009	220	5400	No
15	Dynalifter PSC-1	2009	185	86200	No
16	RA-180/Airship Holland	2009	148	30000	No

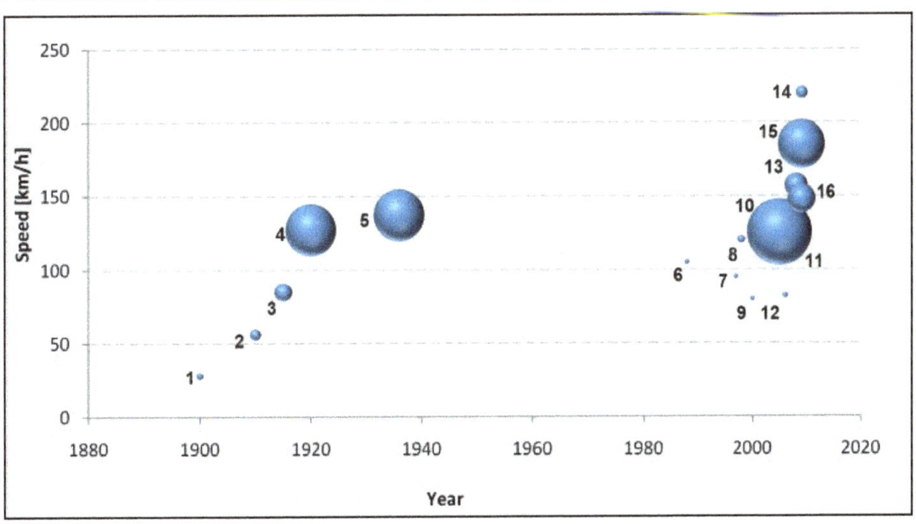

Speed and payload development of airships.

Figure below shows the airship LZ 127 Graf Zeppelin, which circled the earth in 20 days (12 ½ pure driving time). This example should just emphasize the enormous performances those former airships.

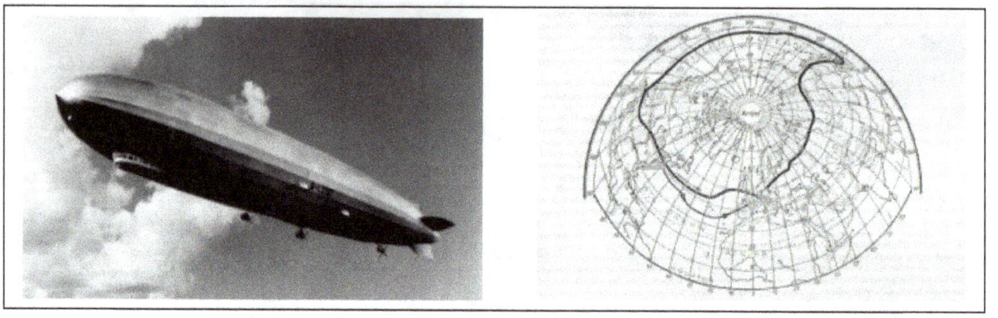

The End of an Airship Era

At the beginning of the Second World War, big airships lost importance. The German military did not find any use and dismantled LZ 127 and LZ130 whereas the US Navy used small blimps for surveillance purposes. Very quickly fixed wing aircraft ruled the skies and became the preferred air transportation system.

Interesting is the aspect that it took about 50 to 60 years since LTA technology could celebrate its revival; however thought for completely different application areas. Since the founding of the Luft Hansa AG in 1926 passenger volume could show a growth of over 338% within the first 12 years. This meant an increase from 56,000 to over 255,000 people within that time span.

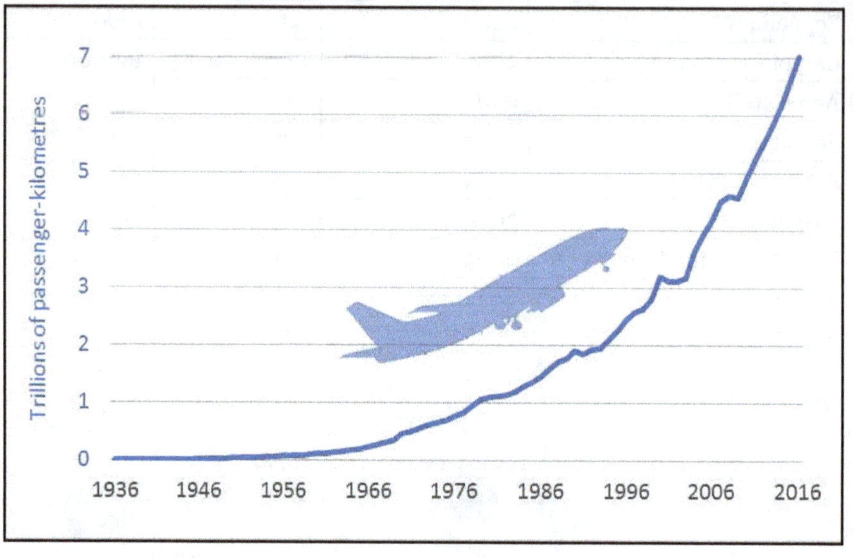

Development of air travel.

Within the same timeframe air cargo grew over 1,080%, an increase from 560 tons to 7,000 tons in 1938. Figure above shows the constant increase in passenger and cargo transportation by fixed wing aircraft. Zeppelin airships instead carried only 18,000 passengers in 28 years. This number almost vanishes in comparison to the 255,000 passengers carried by Luft Hansa in 1937 alone. These numbers clearly show that airships did not have a chance in the emerging mass air transport market, which fixed wing aircraft have been ruling since then.

Free Balloon

A balloon which can be made to ascend by the use of ballast and to descend by the release of gas but which cannot be guided in flight. A balloon is an air-tight bag made out of a light material that can be inflated with air or gas. Toy balloons are available in all kinds of shapes, sizes, and colors to delight children and adults at birthday parties and other festive occasions.

Balloons were first invented in France in the late 18th century. Two papermakers, Jacques and Joseph Montgolfier, discovered that when paper bags are filled with hot air, the bags rise. Quick to realize the potential of this, they began experimenting with balloons of various materials such as paper, cloth, and silk. They made the first public demonstration of a lighter-than-air balloon in June 1783, with a 35-foot (11 m) diameter balloon made of cloth lined with paper. Later that year, Jacques Charles flew a balloon made of silk coated with a rubber varnish and filled with hydrogen, a gas that is lighter than air. These early demonstrations attracted a great deal of excitement, and balloons were soon put to many uses in science, sport, and war.

The rubber toy balloon as we know it today is different from the early balloons in that it is made entirely of rubber. A practical way of making such formed rubber products required several discoveries and inventions. These developments took place gradually over the years since the first rubber factory in the world was established near Paris in 1803.

Natural latex is a mixture of small globules of rubber substance suspended in water (much like milk). When it is exposed to air, heat, or certain chemicals, it coagulates or clots together. The globules of rubber lump together and separate from the watery portion of the latex, eventually forming an elastic, solid material. To improve its strength, resilience, and resistance to hot and cold temperatures, rubber is vulcanized or cured by various methods, such as mixing with certain chemicals or treating with heat.

Raw Materials

Although rubber can be made synthetically, natural latex is preferred for its great elasticity. It can be stretched to seven or eight times its original length and still return to its former shape. Synthetic rubber has not proven to be as elastic and resilient as natural latex.

Raw, natural latex is a white or yellowish opaque liquid, similar in appearance to milk. Latex is the secretion of certain plants, in particular the Hevea tree originally found in Brazil. The most important sources of natural rubber today are plantations in Malaysia and Africa. Producers of rubber must harvest the raw material from these trees, which involves scoring the trees with shallow cuts and letting the sap ooze from the cuts into buckets. The latex is collected in large containers, filtered to remove foreign particles, and mixed with alkali to prevent coagulation. It is then shipped in liquid form to processing centers in different parts of the world.

Latex must be mixed with additives before it can be used in industrial processes. Certain chemicals are mixed in to achieve a desired thickness, rate of drying, and other properties. Other chemicals (collectively known as antidegradants) are added to slow the oxidation and decomposition of the rubber. To give it color, pigments are mixed into the latex. The pigments may be fine metal oxide powders or organic dyes.

Balloons as Flying Machines

A balloon is conceptually the simplest of all flying machines. The balloon is a fabric envelope filled with a gas that is lighter than the surrounding atmosphere. As the entire balloon is less dense than its surroundings, it rises, taking along with it a basket, attached underneath, that carries passengers or payload. The first balloons capable of carrying passengers used hot air to obtain buoyancy and were built by the brothers Josef and Etienne Montgolfier in Annonay, France.

Balloons using the light gas hydrogen for buoyancy were flown less than a month later. They were invented by Professor Jacques Charles and first flown on December 1, 1783. Gas balloons have greater lift and can be flown much longer than hot air, so gas balloons dominated ballooning for the next 200 years. In the 19th century, it was common to use town gas to fill balloons; it was not as light as hydrogen gas, but was much cheaper and readily available.

The third balloon type was invented by Pilâtre de Rozier and is a hybrid of a hot air and a gas balloon. Gas balloons have an advantage of being able to fly for a long time and hot air balloons have an advantage of being able to easily change altitude so the Rozier balloon was a hydrogen balloon with a separate hot air balloon attached. In 1785, Pilâtre de Rozier took off in an attempt to fly across the English Channel but the balloon exploded a half-hour into the flight. This accident earned de Rozier the title "The First to Fly and the First to Die." It wasn't until the 1980s that technology once again allowed the Rozier balloons to become feasible.

A tethered helium balloon gives the public rides to 500 feet above the city of Bristol, England.

Jean-Pierre Blanchard made the first piloted balloon flight in North America on January 9, 1793. Although a balloon has no propulsion system, a degree of directional control is possible through making the balloon rise or sink in altitude to find favorable wind directions. Both the hot-air, and Montgolfière, balloon and the gas balloon are still in common use. Montgolfière balloons are relatively inexpensive as they do not require high-grade materials for their envelopes, and they are popular for balloonist sport activity.

Light gas balloons are predominant in scientific applications, as they are capable of reaching much higher altitudes for much longer periods of time. They are generally filled with helium. Although hydrogen has more lifting power, it is explosive in an atmosphere full of oxygen. With a few exceptions, scientific balloon missions are unmanned. There are two types of light-gas balloons: zero-pressure and superpressure. Zero-pressure balloons are the traditional form of light-gas balloon. They are partially inflated with the light gas before launch, with the gas pressure the same both inside and outside the balloon. As the zero-pressure balloon rises, its gas expands to maintain the zero pressure difference, and the balloon's envelope swells.

At night, the gas in a zero-pressure balloon cools and contracts, causing the balloon to sink. A zero-pressure balloon can only maintain altitude by releasing gas when it goes too high, where the expanding gas can threaten to rupture the envelope, or releasing ballast when it sinks too low. Loss of gas and ballast limits the endurance of zero-pressure balloons to a few days.

A superpressure balloon, in contrast, has a tough and inelastic envelope that is filled with light gas to pressure higher than that of the external atmosphere, and then sealed. The superpressure balloon cannot change size greatly, and so maintains a generally constant volume. The superpressure balloon maintains an altitude of constant density in the atmosphere, and can maintain flight until gas leakage gradually brings it down.

Gas Balloons at the Albuquerque International Balloon Fiesta.

Superpressure balloons offer flight endurance of months, rather than days. In fact, in typical operation an Earth-based superpressure balloon mission is ended by a command from ground control to open the envelope, rather than by natural leakage of gas. For air transport balloons must contain a gas lighter than the surrounding air. There are two types:

- Hot air balloons: Filled with hot air, which by heating becomes lighter than the surrounding air; they have been used to carry human passengers since the 1790s.
- Balloons filled with:
    - Hydrogen: Highly flammable.
    - Helium: Safe if used properly, but very expensive.

Large helium balloons are used as high flying vessels to carry scientific instruments (as do weather balloons), or even human passengers.

Balloons in the Military

The first military use of a balloon was at the Battle of Fleurus in 1794, when L'Entreprenant was used by French Revolutionary troops to watch the movements of the enemy. On the April 2, 1794, an aeronauts corps was created in the French army; however, given the logistical problems linked with the production of hydrogen on the battlefield (it required constructing ovens and pouring water on white-hot iron), the corps was disbanded in 1799.

American Civil War

The first major-scale use of balloons in the military occurred during the American Civil War with the Union Army Balloon Corps established and organized by Prof. Thaddeus S. C. Lowe in the summer of 1861. The balloons were inflated with coke gas from municipal services and then walked out to the battlefield, an arduous and inefficient operation as the balloons had to be returned to the city every four days for re-inflation. Eventually hydrogen gas generators, a compact system of tanks and copper plumbing, were constructed which converted the combining of iron filings and sulphuric acid to hydrogen. The generators were easily transported with the uninflated balloons to the field on a standard buckboard. In all Lowe built seven balloons that were fit for military service.

The first application thought useful for balloons was map making from aerial vantage points, thus Lowe's first assignment was with the Topographical Engineers. General Irvin McDowell, commander of the Grand Army of the Potomac, realized their value in aerial reconnaissance and had Lowe, who at the time was using his personal balloon the Enterprise, called up to the First Battle of Bull Run. In a later exercise, Lowe was called to act as a Forward Artillery Observer (FAO) from which aerial station he was able to direct artillery fire by a set order of flag signals, from an unseen position, onto a Confederate encampment. FAO is still used today from either ground or aerial positions.

Lowe's first military balloon, the Eagle was ready by October 1st, 1861. It was called into service immediately to be towed to Lewinsville, Virginia, without any gas generator which took longer to build. The trip began after inflation in Washington and turned into a 12-mile, 12-hour excursion that was upended by a gale force wind which ripped the aerostat from its netting and sent it sailing to the coast. Balloon activities were suspended until all balloons and gas generators were completed.

With his ability to inflate balloons from remote stations, Lowe, his new balloon the Washington and two gas generators were loaded onto a converted coal barge the George Washington Parke Custis. As he was towed down the Potomac, Lowe was able to ascend and observe the battlefield as it moved inward on the heavily forested peninsula. This would be the military's first claim of an aircraft carrier. The Union Army Balloon Corps enjoyed more success in the battles of the Peninsula Campaign than the Army of the Potomac it sought to support. The general military attitude toward the use of balloons deteriorated, and by August 1863 the Balloon Corps was disbanded.

Confederate Army Uses

The Confederate Army made use of balloons, but they were gravely hampered by supplies due to the embargoes. They were forced to fashion their balloons from gaily colored silk dress-making material, and their use was limited by the infrequent supply of gas in Richmond, Virginia. By the summer of 1863, all balloon reconnaissance of the Civil War had ceased.

In Other Countries

In Britain during July 1863 experimental balloon ascents for reconnaissance purposes were conducted by the Royal Engineers on behalf of the British Army, but although the experiments were successful it was considered not worth pursuing further because it was too expensive. However by 1888 a School of Ballooning was established at Chatham, Medway Kent. It moved to Stanhope Lines, Aldershot in 1890 when a balloon section and depot were formed as permanent units of the Royal Engineers establishment.

Balloons were used by the Royal Engineers for reconnaissance and observation purposes during the Bechuanaland Expedition, the Sudan Expedition and during the Anglo Boer War. On 5 October 1907 Colonel John Capper (late Royal Engineers) and team flew the military airship Nulli Secundus from Farnborough round At Paul's Cathedral in London and back with a view to raising public interest.

Hydrogen-filled balloons were also widely used during World War I to detect enemy troop movements and to direct artillery fire. Observers phoned their reports to officers on the ground who then relayed the information to those who needed it.

Because artillery was such an important factor in World War I, balloons were frequent targets of opposing aircraft. Though balloon companies were protected by antiaircraft guns and patrolling fighters, casualties were frequently heavy. One reason for this was the hydrogen that filled the balloons was highly flammable, and planes assigned to attack enemy balloons were often equipped with incendiary bullets, for the purpose of igniting the balloon.

The Aeronaut Badge was established by the United States Army in World War I to denote service members who were qualified balloon pilots. Observation balloons were retained well after the Great War, being used in the Russo-Finnish conflicts.

The Japanese launched thousands of balloon bombs to the US and Canada, carried in the jet stream. The British used balloons to carry incendiary devices to Germany between 1942 and 1944.

In Space

Balloons have a number of advantages for planetary exploration. They can be made light in weight and are potentially relatively inexpensive. They can cover a great deal of territory, and their view from a height gives them the ability to examine wide swathes of terrain with far more detail than would be available from an orbiting satellite. For exploratory missions, their relative lack of directional control is not a major obstacle as there is generally no need to direct them to a specific location.

In June 1985, the Russian space probes Vega 1 and Vega 2 each released an aerobot (balloon-like robot) into the atmosphere of Venus. Signals from the first balloon were received for only 56 minutes, but those from the second arrived for nearly two Earth days until its batteries ran down.

Balloons such as the Echo satellite are launched with a rocket. They are not the typical balloon, but simply large deployable structures.

In Medicine

Angioplasty is a surgical procedure in which very small balloons are inserted into blocked or partially blocked blood vessels near the heart. Once in place, the balloon can be inflated to clear or compress arterial plaque, and to stretch the walls of the vein. A small stent can be inserted in its place to keep the vessel open after the balloon's removal.

Certain cathcters have balloons at their tip to keep them from slipping out, for example the balloon of a Foley catheter is insufflated when the catheter is inserted into the urinary bladder and secures its position.

Records

- Manned Balloon: The altitude record for manned balloons is 34668 metres. It was made by Malcolm D. Ross and Victor E. Prather over the Gulf of Mexico in 1961.
- Unmanned Balloon: The altitude record for unmanned balloons is (1991 edition of Guinness Book) 51.8 kilometres. The vehicle was a Winzen-Balloon with a volume of 1.35 million cubic metres, which was launched in October 1972 in Chico, California, USA. This is the greatest altitude ever reached by a flying object requiring the surrounding air. Higher altitudes can only be reached by ballistic vehicles such as rockets, rocket planes or projectiles.

Manufacturing of Balloons

The process of making a toy balloon involves dipping a mold into liquid latex. The mold, or form, is shaped like a deflated balloon. The earliest balloon forms were disposable, made from cardboard attached to dowels. Modern forms are reusable and usually made from stainless steel, aluminum, or porcelain. The forms must be smooth and polished. A number of such forms are attached upside down to a board or rack. The boards are moved mechanically from one station to another in the factory.

To be efficient in terms of cost and number of balloons produced, balloon manufacture has become a highly automated, continuous loop process. Balloons are made in batches, all of the same color and size, since changing the color and form is time-consuming and requires manual intervention. Manual intervention is usually only needed for setting up a run and then later for packaging the finished product, and for dealing with occasional mechanical problems that may arise.

Preparing the Latex

- Prior to its use, the latex may need to be colored. This involves mixing a pigment into the latex. It may be done at the balloon factory, or the balloon maker may purchase already-pigmented latex from a supplier.

- The latex must be poured into tanks into which the forms will be dipped. The tanks are kept at a certain temperature and may include stirring mechanisms to keep the latex circulating to avoid settling.

Dipping the Forms

- The balloon forms are first heated, then immersed in a tank of coagulant solution for a few seconds. When the forms are immersed in the liquid latex, the coagulant will cause the rubber to gel in a thin sheet around the forms. A commonly used coagulant solution is a mixture of water, a calcium-based salt, soap, and talc powder. The salt is the actual coagulant; the soap helps the latex spread in an even film, and the talc helps ease the removal of the rubber from the forms in a later step.
- The forms are heated to a temperature between 100 °F (38 °C) and 200°F (93 °C), and then immersed in a tank of colored latex. The coagulant causes the latex to coat the forms. The longer the forms are left in the tank, the thicker the coating that sticks to them. For balloons, a very thin layer of latex is desired, so the forms are immersed only for a few seconds. The forms must be inserted and removed at carefully controlled speeds to avoid trapping air bubbles and to achieve an even, thin coating.

Making the Ring

A lip is formed on the neck of the balloon by rolling the edges of the rubber using brushes or rollers. This creates the ring seen around the opening of the balloon.

Removing Excess Coagulant

Next, the forms are immersed in a tank of leaching solution (often plain water) to dissolve and leach away excess coagulant from the rubber.

Curing the Rubber

The rubber on the forms must be dried and cured. The method used varies among manufacturers. Some balloon makers use a latex that already contains a vulcanizing agent, in which case the rubber is dried at a moderate temperature. Other makers induce vulcanization by putting the rubber-coated forms into an oven and curing for as long as an hour.

Removing the Balloons

- The balloons are then mechanically removed from their forms. One approach is to blow them off using a spray of water or air and collecting the balloons in a basket or net.
- If the balloons are removed using a spray of water, they are next placed in a centrifuge, where excess water is removed by spinning the balloons around at high speed.
- The balloons are then dried in large tumble dryers.

Printing and Packaging

- Next, the balloons may either be packaged, or first printed and then packaged. If they are packaged directly, they are moved on a conveyor belt past a counting device and placed into bags. When an appropriate number of balloons has been placed in each bag, the bags are sealed.
- Printing designs on balloons, such as logos or faces, actually involves several steps. First, the balloons must be inflated in order to allow even printing. This requires a worker to manually place each balloon on the inflating device. Next, a pattern is carefully printed on each balloon. Finally, the balloons are removed and passed on to the packaging stage.

Quality Control

The balloon manufacturing environment must be strictly controlled in order to achieve high quality and consistency. Throughout the manufacturing process, computer-based instrumentation records and controls air humidity, air temperature, latex tank temperature, the temperature in the ovens, dryers, and other parameters.

The latex and other chemicals used in the process must be carefully formulated for specific properties, and carefully maintained. For example, the latex must have certain viscosity and speed of drying. The tanks in which it is held must have devices to keep the latex circulating to avoid forming a "skin," and to prevent ingredients from settling.

By Products/Waste

It is in the manufacturers' best interests to waste as little rubber as possible because the cost of latex is high compared to the selling price of individual balloons. Balloon makers also reclaim much of the coagulant that ends up in the leaching solution. Unfortunately, what is not reclaimed ends up as liquid waste in the environment. The amount of chemical waste that can be released by a factory is regulated by government laws. Balloons also result in some waste after they are manufactured because they are invariably thrown away after they deflate or pop. However, because latex is natural, it eventually breaks down into other substances.

Safety Concerns

Toy balloons can be a source of joy, but they can also be unexpectedly hazardous. Latex balloons may also end up in water, where they eventually lose their color and can resemble jellyfish. Sea animals such as whales and turtles have attempted to eat them and have died because the latex clogs their digestive systems.

Heavier-than-Air Systems

This type of aircraft must have a power source to provide the thrust necessary to obtain lift. Simple heavier-than-air craft include kites. These are usually a flat-surfaced structure, often with a stabilizing "tail," attached by a bridle to a string that is held in place on the ground. Lift is provided by the reaction of the string-restrained surface to the wind. Another type of unmanned aircraft is the unmanned aerial vehicle (UAV), commonly called drones or sometimes remotely piloted vehicles (RPVs). These aircraft are radio-controlled from the air or the ground and are used for scientific and military purposes.

Israeli Aircraft Industries Searcher, a reconnaissance uncrewed aerial vehicle, at Tel Nof air base, Israel.

Unpowered manned heavier-than-air vehicles must be launched to obtain lift. These include hang gliders, gliders, and sailplanes. Hang gliders are aircraft of various configurations in which the pilot is suspended beneath the (usually fabric) wing to provide stability and control. They are normally launched from a high point. In the hands of an experienced pilot, hang gliders are capable of soaring (using rising air columns to obtain upward gliding movement).

Glider.

Gliders are usually used for flight training and have the capability to fly reasonable distances when they are catapulted or towed into the air, but they lack the dynamic sophistication of sailplanes. These sophisticated unpowered craft have wings of unusually high aspect ratio (that is, a long wing span in proportion to wing width). Most sailplanes are towed to launch altitude, although some employ small, retractable auxiliary engines. They are able to use thermals (currents more buoyant than the surrounding air, usually caused by higher temperature) and orographic lift to climb to higher altitude and to glide for great distances. Orographic lift results from the mechanical effect of wind blowing against a terrain feature such as a cliff. The force of the wind is deflected upward by the face of the terrain, resulting in a rising current of air.

Ultralights, which were originally merely hang gliders adapted for power by the installation of small engines similar to those used in chain saws, have matured into specially designed aircraft of very low weight and power but with flying qualities similar to conventional light aircraft. They are intended primarily for pleasure flying, although advanced models are now used for training, police patrol, and other work, including a proposed use in combat.

Whooping cranes following an ultralight aircraft.

Experimental craft have been designed to make use of human and solar power. These are very lightweight, sophisticated aircraft, designed with heavy reliance on computers and using the most modern materials. Solar-powered aircraft are similar to human-powered types, except that they use solar panels to convert the Sun's energy directly to power an electric motor.

Helicopters

The wings of the airplane create a lift force when they move through the air. During flight, there are four forces acting on the helicopter or airplane and those are lift, drag, thrust, and weight. In order to make the wings to move through the air, of course, the plane itself has to move. A helicopter works by having its wings move through the air while the body stays still. The helicopter's wings are called Main Rotor Blades. The shape and the angle of the blades move through the air will determine how much Lift force is created. After the helicopter lifted off the ground, the pilot can tilt the blades, causing the helicopter to tip forward or backward or sideward.

Nomenclature and Technical Term

- Bernoulli's principle states that as the air velocity increases, the pressure decreases; and as the velocity decreases, the pressure increases.
- Airfoil is technically defined as any surface, such as an airplane aileron, elevator, rudder, wing, main rotor blades, or tail rotor blades designed to obtain reaction from the air through which it moves.

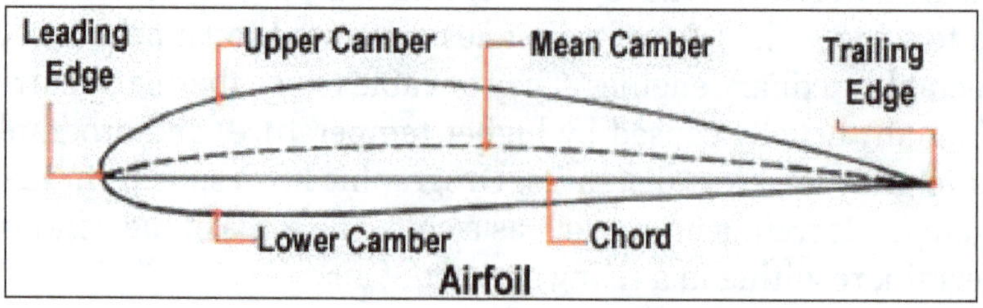

- Angle of Attack is the acute angle measured between the chord of an airfoil and the relative wind.

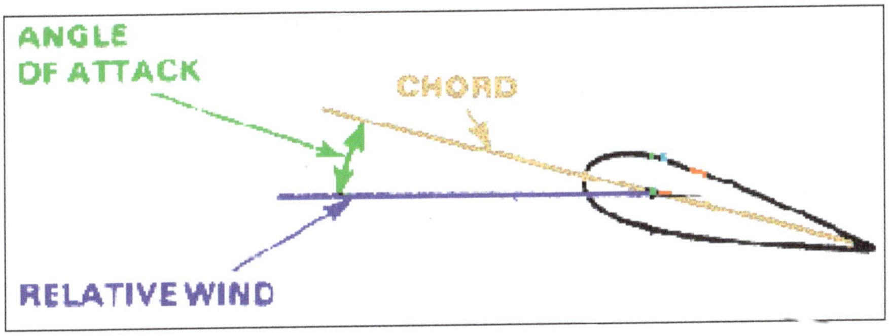

- Angle of Incidence is the acute angle between the wing's chord line and the longitudinal axis of the airplane. (Usually manufacturer had built the aircraft with the wing has some degrees to the horizontal plane or airplane longitudinal axis).
- Blades of the helicopter are airfoils with a very high aspect ratio (length to chord). The angle of incidence is adjusted by means of the control from pilots.

The main rotor of the helicopter may have two, three, four, five or six blades, depending upon the design. The main rotor blades are hinged to the rotor head in such a manner that they have limited movement up and down and also they can change the pitch (angle of incidence). The controls for the main rotor are called Collective and Cyclic Controls.

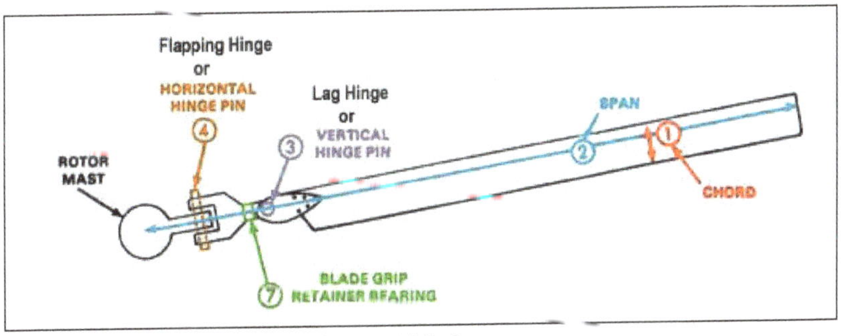

Main rotor blade system.

The tail rotor is small blades may have two or four blades and mounted on the tail of the helicopter, it rotates in the vertical plane. The tail rotor is controlled by the rudder pedals. Its pitch can be changed as required to turn the helicopter in the direction desired.

- Blade Root: The inner end of the blades where the rotors connect to the blade gripos.

- Blade Grips: Large attaching points where the rotor blade connects to the hub.
- Rotor Hub: Sit on top of the mast, and connects the rotor blades to the control tubes.
- Main Rotor Mast: Rotating shaft from the transmission which connects the main rotor blades to helicopter fuselage.

- Pitch Change Horn: To converts control tube movement to blade pitch. Control tube is a push-pull tubes that change the pitch of the rotor blades through the pitch changing horn.
- Swash Plate Assembly: The swash plate assembly consists of two primary elements through which the rotor mast passes. One element is a disc, linked to the cyclic pitch control. This disc is capable of tilting in any direction but does not rotate as the rotor rotates. This non-rotating disc, often refered to as the Stationary Star is attached by a bearing surface to a second disc, often refered to as the Rotating Star which turns with rotor and linked to the rotor blade pitch horns.

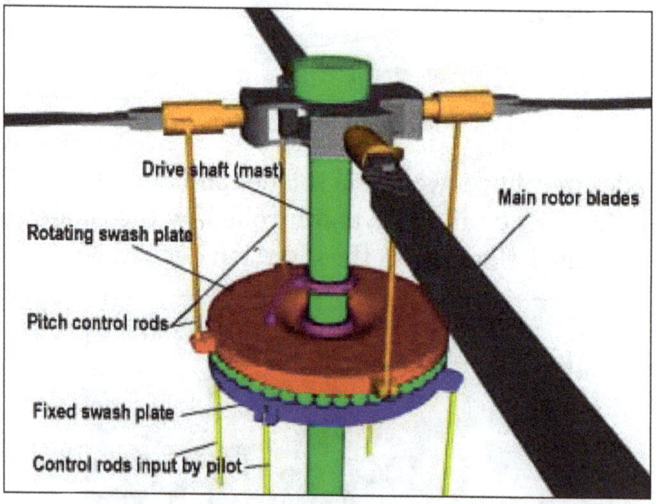

- Transmission: The transmission system transmits engine power to the main rotor, tail rotor, generator and other accessories. The engine is operated at a relative high speed while the main rotor turns at a much lower speed. This speed reduction is accomplished through reduction gears in the Transmission System.
- Lift: is produced by a lower pressure created on the upper surface of an airplane's wings

compared to the pressure on the wing's lower surfaces, causing the wing to be LIFTED upward. The special shape of the airplane wing (airfoil) is designed so that air flowing over it will have to travel a greater distance and faster resulting in a lower pressure area thus lifting the wing upward. Lift is that force which opposes the force of gravity (or weight).

Lift depends upon: (1) shape of the airfoil (2) the angle of attack (3) the area of the surface exposed to the airstream (4) the square of the air speed (5) the air density.

Lift equation:

$$L = C_L \times d \times \frac{V^2}{2} \times A$$

L = Lift

C_L = Lift coefficient

d = Density of air

V = Velocity of air

A = Wing area

- Relative Wind is the direction of the airflow with respect to an airfoil or to the rotor blades.
- Pitch Angle: The rotor blade pitch angle is the acute angle between the blade chord line and the rotor plane of rotation. (You may understand as the angle of incidence). This pitch angle can be varied by the pilot through the use of cockpit controls.

Main Rotor

The lifting force is produced by the main rotor. As they spin in the air and produced the lift. Each blade produces an equal share of the lifting force. The weight of a helicopter is divided evenly between the rotor blades on the main rotor system. If a helicopter weight 4000 lbs and it has two blades, then each blade must be able to support 2000 lbs. In addition to the static weight of helicopter, each blade must be accept dynamic load as well. For example, if a helicopter pull up in a 1.5g manouver (1.5 time the gravity force), then the effective weight of helicopter will be 1.5 time of static helicopter weight or 6000 lbs. due to gravitational pull.

Tail Rotor

The tail rotor is very important. If you spin a rotor with an engine, the rotor will rotate, but the engine and helicopter body will tend to rotate in opposite direction to the rotor. This is called Torque reaction. Newton's third law of motion states, "to every action there is an equal and opposite reaction". The tail rotor is used to compensates for this torque and hold the helicopter straight. On twin-rotors helicopter, the rotors spin in opposite directions, so their reactions cancel each other.

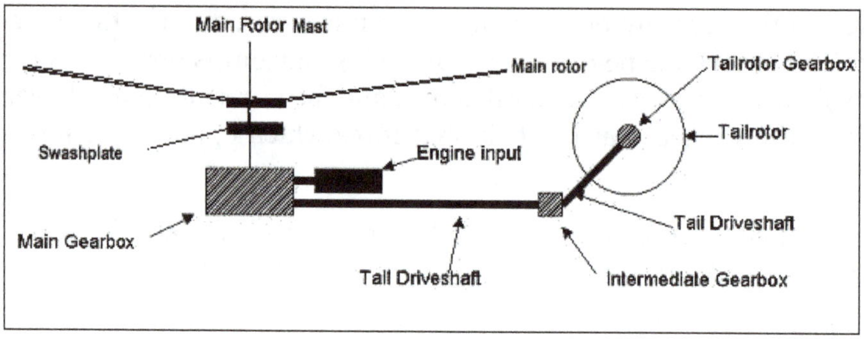

Typical helicopter drive train.

The tail rotor in normally linked to the main rotor via a system of drives hafts and gearboxes, that means if you turn the main rotor, the tail rotor is also turn. Most helicopter have a ratio of 3:1 to 6:1. That is, if main rotor turn one rotation, the tail rotor will turn 3 revolutions (for 3:1) or 6 revolutions (for 6:1). In most helicopter the engine turns a shaft that connected to an input quill in the transmission gearbox. the main rotor mast out to the top and tail rotor drive shafts out to the tail from the transmission gear box.

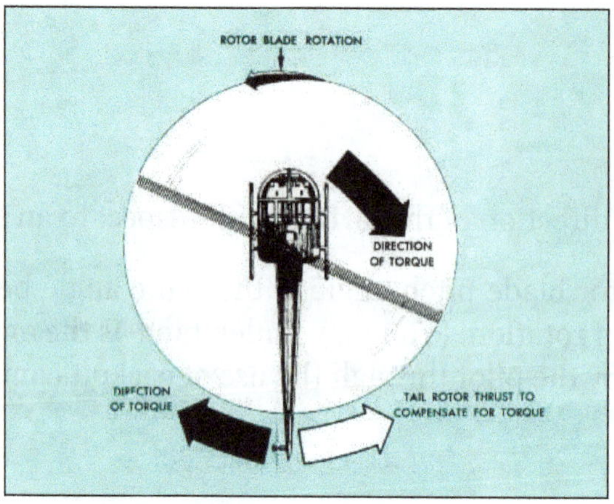

Torque effect and tail rotor thrust compensate.

Dissymmetry of Lift

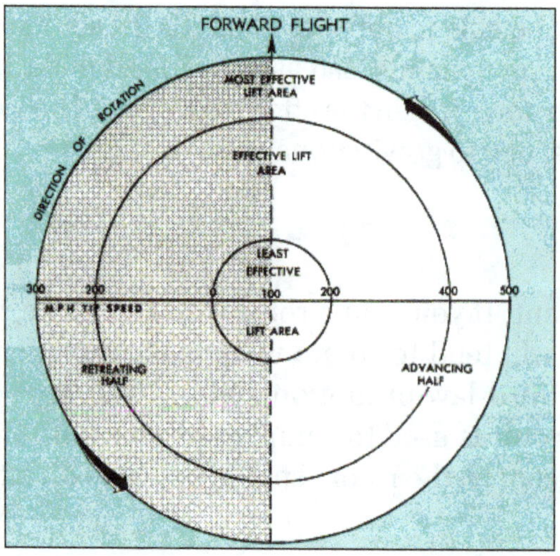

All rotor systems are subject to Dissymmetry of Lift in forward flight. At a hover, the lift is equal across the entire rotor disk. As the helicopter gain air speed, the advancing blade develops greater lift because of the increased airspeed and the retreating blade will produce less lift, this will cause the helicopter to roll (for example: if rotor speed = 400 km/hr, helicopter move forward=100 km/hr then advancing blade will have speed=500 km/hr but the retreating blade will has moving speed of only 300 kr/hr). This has to be compensated for in some way.

Blade Flapping

Dissymmetry of lift is compensated for by Blade flapping. Because of the increased airspeed and lift on the advancing blade will cause the blade to flap up and decreasing the angle of attack. The decreased lift on the retreating blade will cause the blade to flap down and increasing the angle of attack. The combination of decreased angle of attack on the advancing blade and increased angle of attack on the retreating blade through blade flapping action tends to equalize the lift over the two halves of the rotor disc.

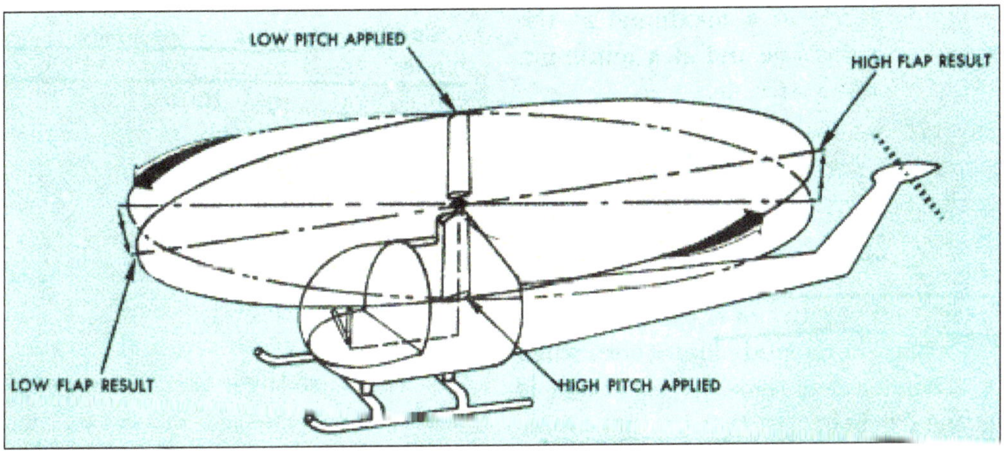

Relationship between Swash Plate and Controls

- Swash Plate Assembly: The swash plate assembly consists of two primary elements through which the rotor mast passes. One element is a disc, linked to the cyclic pitch control. This disc is capable of tilting in any direction but does not rotate as the rotor rotates. This non-rotating disc, often refered to as the Stationary Star is attached by a bearing surface to a second disc, often refered to as the Rotating Star which turns with rotor and linked to the rotor blade pitch horns.
- The Collective Control: When pilot raises the collective control or pull collective control up, the collective control will raises the entire swash plate assembly as a unit. This has effect to the blades by changing the pitch of all blades simultaneously. This causes to increase angle of attack and give more lift.
- The Cyclic Control: The cyclic control will push one side of the swashplate assembly up or down. This has the effect to the rotor head system because the cyclic control or cyclic stick controls the angle of the main rotor by angling the rotor head to which all the blades are attached .This cause the helicopter to move left or right, forward or backward.

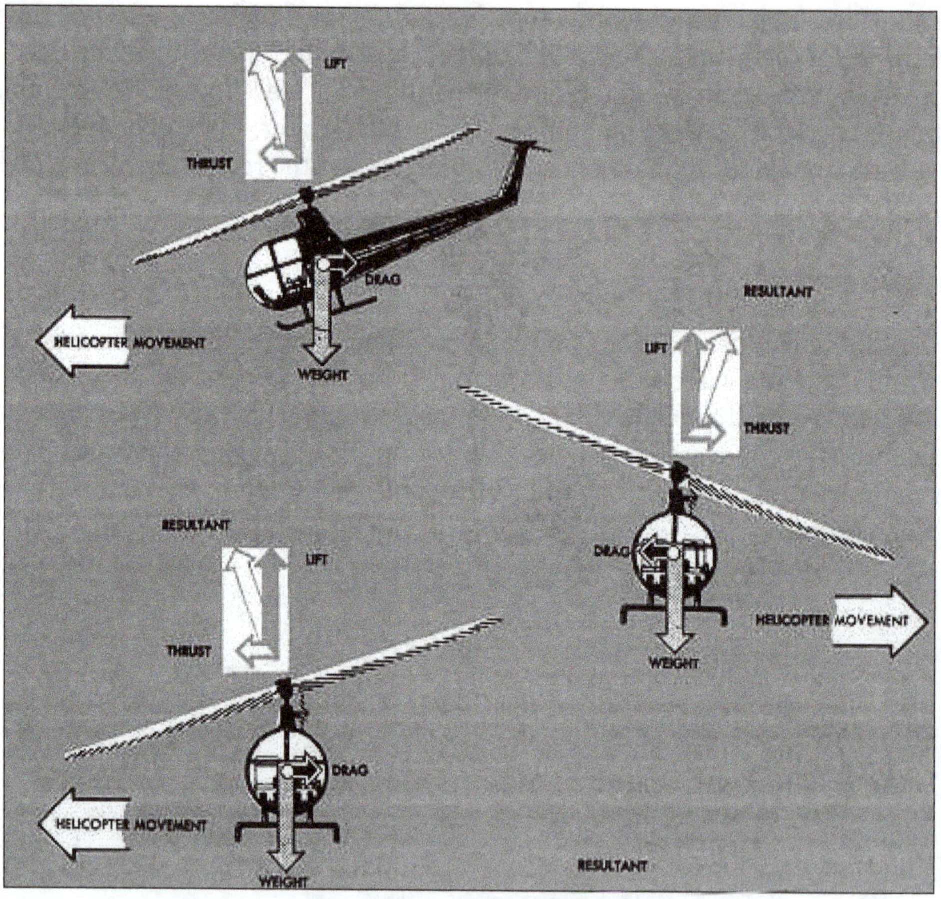

Forces acting and movement.

Anti-torque Pedals

The Thrust produced by the auxiliary (tail) rotor is governed by the position of anti-torque pedals. These are not rudder pedals, although they are in the same place as rudder pedals on an airplane. They are linked to a pitch change mechanism in the tail rotor gear box to permit the pilot to increase the pitch of the tail rotor blades. The primary purpose of the tail rotor and its controls is to counteract the torque effect of the main rotor.

Gyroplanes

Gyroplane is the official name for an aircraft that uses a free-turning rotary wing. They are also known as gyrocopters, gyros and sometimes, autogyros. It is defined as 'a heavier-than-air aircraft that is supported in flight by the reaction of the air on rotors that are not power-driven, but rotate when the aircraft is moving because of the action of the air; and that has a power-driven propulsion system that is independent of the rotors.'

A gyro looks like a helicopter to the layperson but differs in two fundamental ways. The rotor blades are turned not by an engine, but by the gyro's forward motion through the air. And that forward motion comes not from tilting the main rotor disc as a helicopter does, but from a separate propeller. A gyro's rotor is in constant autorotation, just as a helicopter's is in an unpowered landing.

Complicating the definition is the fact that some gyros pre-rotate their rotors before take-off with a driveshaft clutched to the propeller engine, or an electric motor. But this is only to allow a shorter take-off and never used in flight.

Gyroplanes have somewhat different flying and performance characteristics to fixed wing aircraft. They have high drag and are therefore relatively slow, with typical cruise speeds of between 50 and 60 kt. The gyro's advantage is not top speed but its ability to fly slowly, which is why under the name autogyro they were developed in the 1930s as a precursor to the helicopter. Unlike any fixed-wing aircraft they have the sometimes-handy ability to safely make a vertical descent to wash off altitude quickly.

Gyros cannot stall because the rotational speed of their rotors, and hence their lift, is not dependent on forward speed. At zero forward airspeed a gyro will slowly descend, to a survivable landing. However, gyros can lose rotor speed if the rotor disc angle becomes too shallow. In normal forward flight the gyro rotor disc tilts backwards, but as all properly trained gyro pilots know, the disc comes closer to horizontal as forward speed increases, (in the same way a fixed-wing's angle of attack decreases with speed). This increases the risk of a power push-over and loss of rotor speed as forward speed rises.

With its excess of lift the gyro can execute a steep turn at any time—it doesn't need to be travelling fast to do this. However, the flip side of its aerodynamic coin is a poor glide ratio, at best in the vicinity of 4:1 in ideal conditions and no wind. But when the wind blows, gyros can handle it. Many designs are able to handle wind strengths exceeding 35 kt safely.

Gyros are often described as short take-off and landing aircraft. This is not always true; take-off distance depends on ambient conditions (like any other aircraft), particularly headwind strength, and the amount of pre-rotation applied by the pilot before the gyro starts its roll. Depending on wind strength, gyros can come to a halt within 30 metres or less of touch-down point. At this stage the gyro's rotors are still rotating near flying speed so a gyro pilot will pause with the stick fully back and then move the stick forward, tilting the rotor disc forward to allow the rotor speed to decay before leaving the runway. This pause is typically fewer than 30 seconds.

Gyros are classified as an ultralight with a maximum flying altitude of 500 ft AGL unless the pilot holds a specific endorsement. Gyros may also legally operate at a minimum altitude of 300 ft AGL and even lower over private property with the permission of the landowner.

A standard gyro circuit is flown at 500 ft AGL. Gyros fly their circuit legs close to the active runway for two reasons. Firstly, they do this to compensate for their poor glide ratio in case of engine failure; and secondly, to improve the likelihood of faster aircraft seeing the gyro in the circuit.

Gyro pilots are trained, wherever possible, to use idle power on approach to ensure they become well practised in simulated engine-out approaches. This means gyros typically fly a short final with a steep approach angle. This technique can lead to potential traffic conflict with faster fixed wing aircraft that may have to do a last minute go-around when they finally see the gyro ahead and below them. But these potential conflicts can be easily sorted out by sensible use of radio to advise and coordinate between aircraft. Fixed-wing pilots waiting to take off also occasionally fail to look up as well as out, thus forcing a landing gyro to take last-minute evasive action to avoid a possible runway incursion.

References

- Aircraft: cs.mcgill.ca, Retrieved 26, May 2020
- Application-of-Lighter-Than-Air-Platforms-for-Power-Beaming-Generation-and-Communications-224266478: researchgate.net, Retrieved 14, January 2020
- Types-of-aircraft, airplane, technology-64156: britannica.com, Retrieved 07, July 2020
- Heli-principle, helicopter: thaitechnics.com, Retrieved 28, April 2020
- Sharing-the-sky-gyroplanes: flightsafetyaustralia.com, Retrieved 14, February 2020

3

Manufacturing of Aircraft

The manufacturing process of aircrafts includes the design, fabrication, assembly, testing and inspection of different components of the aircraft. This chapter has been carefully written to provide an easy understanding of the varied facets of aircraft manufacturing process such as aircraft design, flight control system, aviation and aeronautics.

Aircraft Design Process

The aircraft design process is the engineering design process by which aircraft are designed. These depend on many factors such as customer and manufacturer demand, safety protocols, physical and economic constraints etc. For some types of aircraft the design process is regulated by national airworthiness authorities. This article deals with powered aircraft such as airplanes and helicopter designs.

AST model in wind tunnel.

Aircraft design is a compromise between many competing factors and constraints and accounts for existing designs and market requirements to produce the best aircraft.

Design Constraints

Purpose

The design process starts with the aircraft's intended purpose. Commercial airliners are designed for carrying a passenger or cargo payload, long range and greater fuel efficiency where as fighter jets are designed to perform high speed maneuvers and provide close support to ground troops. Some aircraft have specific missions, for instance, amphibious airplanes have a unique design that allows them to operate from both land and water, some fighters, like the Harrier Jump Jet, have VTOL (Vertical Take-off and Landing) ability, helicopters have the ability to hover over an area for a period of time.

The purpose may be to fit a specific requirement, e.g. as in the historical case of a British Air Ministry specification, or fill a perceived "gap in the market"; that is, a class or design of aircraft which does not yet exist, but for which there would be significant demand.

Aircraft Regulations

Another important factor that influences the design of the aircraft are the regulations put forth by national aviation airworthiness authorities.

Airports may also impose limits on aircraft, for instance, the maximum wingspan allowed for a conventional aircraft is 80 m to prevent collisions between aircraft while taxiing.

Financial Factors and Market

Budget limitations, market requirements and competition set constraints on the design process and comprise the non-technical influences on aircraft design along with environmental factors. Competition leads to companies striving for better efficiency in the design without compromising performance and incorporating new techniques and technology.

Environmental Factors

An increase in the number of aircraft also means greater carbon emissions. Environmental scientists have voiced concern over the main kinds of pollution associated with aircraft, mainly noise and emissions. Aircraft engines have been historically notorious for creating noise pollution and the expansion of airways over already congested and polluted cities have drawn heavy criticism, making it necessary to have environmental policies for aircraft noise. Noise also arises from the airframe, where the airflow directions are changed. Improved noise regulations have forced designers to create quieter engines and airframes. Emissions from aircraft include particulates, carbon dioxide (CO_2), Sulfur dioxide(SO_2), Carbon monoxide (CO), various oxides of nitrates and unburnt hydrocarbons. To combat the pollution, ICAO set recommendations in 1981 to control aircraft emissions. Newer, environmentally friendly fuels have been developed and the use of recyclable materials in manufacturing have helped reduce the ecological impact due to aircraft. Environmental limitations also affect airfield compatibility. Airports around the world have been built to suit the topography of the particular region. Space limitations, pavement design, runway end safety areas and the unique location of airport are some of the airport factors that influence aircraft design. However changes in aircraft design also influence airfield design as well, for instance, the recent introduction of new large aircraft (NLAs) such as the superjumbo Airbus A380, have led to airports worldwide redesigning their facilities to accommodate its large size and service requirements.

Safety

The high speeds, fuel tanks, atmospheric conditions at cruise altitudes, natural hazards (thunderstorms, hail and bird strikes) and human error are some of the many hazards that pose a threat to air travel.

Airworthiness is the standard by which aircraft are determined fit to fly. The responsibility for airworthiness lies with national aviation regulatory bodies, manufacturers, as well as owners and operators.

The International Civil Aviation Organization sets international standards and recommended practices for national authorities to base their regulations on The national regulatory authorities set standards for airworthiness, issue certificates to manufacturers and operators and the standards of personnel training. Every country has its own regulatory body such as the Federal Aviation Authority in USA, DGCA (Directorate General of Civil Aviation) in India, etc.

The aircraft manufacturer makes sure that the aircraft meets existing design standards, defines the operating limitations and maintenance schedules and provides support and maintenance throughout the operational life of the aircraft. The aviation operators include the passenger and cargo airliners, air forces and owners of private aircraft. They agree to comply with the regulations set by the regulatory bodies, understand the limitations of the aircraft as specified by the manufacturer, report defects and assist the manufacturers in keeping up the airworthiness standards.

Most of the design criticisms these days are built on crashworthiness. Even with the greatest attention to airworthiness, accidents still occur. Crashworthiness is the qualitative evaluation of how aircraft survive an accident. The main objective is to protect the passengers or valuable cargo from the damage caused by an accident. In the case of airliners the stressed skin of the pressurized fuselage provides this feature, but in the event of a nose or tail impact, large bending moments build all the way through the fuselage, causing fractures in the shell, causing the fuselage to break up into smaller sections. So the passenger aircraft are designed in such a way that seating arrangements are away from areas likely to be intruded in an accident, such as near a propeller, engine nacelle undercarriage etc. The interior of the cabin is also fitted with safety features such as oxygen masks that drop down in the event of loss of cabin pressure, lockable luggage compartments, safety belts, lifejackets, emergency doors and luminous floor strips. Aircraft are sometimes designed with emergency water landing in mind, for instance the Airbus A330 has a 'ditching' switch that closes valves and openings beneath the aircraft slowing the ingress of water.

Design Optimization

Aircraft designers normally rough-out the initial design with consideration of all the constraints on their design. Historically design teams used to be small, usually headed by a Chief Designer who knows all the design requirements and objectives and coordinated the team accordingly. As time progressed, the complexity of military and airline aircraft also grew. Modern military and airline design projects are of such a large scale that every design aspect is tackled by different teams and then brought together. In general aviation a large number of light aircraft are designed and built by amateur hobbyists and enthusiasts.

Computer-aided Design of Aircraft

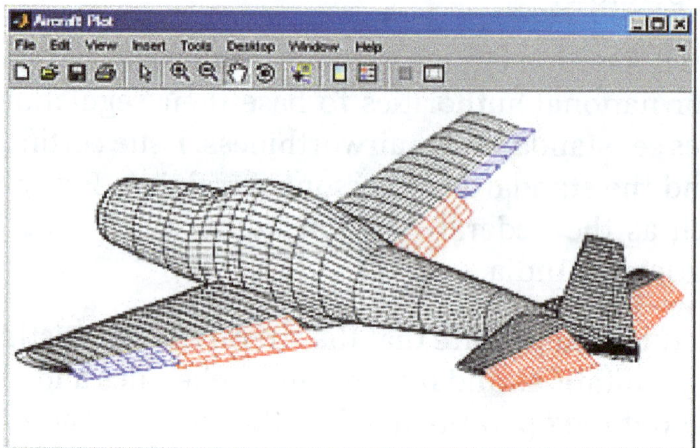

The external surfaces of an aircraft modelled in MATLAB.

In the early years of aircraft design, designers generally used analytical theory to do the various engineering calculations that go into the design process along with a lot of experimentation. These calculations were labour-intensive and time-consuming. In the 1940s, several engineers started looking for ways to automate and simplify the calculation process and many relations and semi-empirical formulas were developed. Even after simplification, the calculations continued to be extensive. With the invention of the computer, engineers realized that a majority of the calculations could be automated, but the lack of design visualization and the huge amount of experimentation involved kept the field of aircraft design stagnant. With the rise of programming languages, engineers could now write programs that were tailored to design an aircraft. Originally this was done with mainframe computers and used low-level programming languages that required the user to be fluent in the language and know the architecture of the computer. With the introduction of personal computers, design programs began employing a more user-friendly approach.

Design Aspects

The main aspects of aircraft design are:

1. Aerodynamics
2. Propulsion
3. Controls
4. Mass
5. Structure

All aircraft designs involve compromises of these factors to achieve the design mission.

Wing Design

The wings of a fixed wing aircraft provide the necessary lift for take-off and cruise flight. Wing geometry affects every aspect of an aircraft's flight. The wing area will usually be dictated by aircraft performance requirements (e.g. field length) but the overall shape of the planform and other detail aspects may be influenced by wing layout factors. The wing can be mounted to the fuselage in high, low and middle positions. The wing design depends on many parameters such as selection of aspect ratio, taper ratio, sweepback angle, thickness ratio, section profile, washout and dihedral. The cross-sectional shape of the wing is its airfoil. The construction of the wing starts with the rib which defines the airfoil shape. Ribs can be made of wood, metal, plastic or even composites.

Fuselage

The fuselage is the part of the aircraft that contains the cockpit, passenger cabin or cargo hold.

Propulsion

Aircraft engine.

Aircraft propulsion may be achieved by specially designed aircraft engines, adapted auto, motorcycle or snowmobile engines, electric engines or even human muscle power. The main parameters of engine design are:

- Maximum engine thrust available.
- Fuel consumption.
- Engine mass.
- Engine geometry.

The thrust provided by the engine must balance the drag at cruise speed and be greater than the drag to allow acceleration. The engine requirement varies with the type of aircraft. For instance, commercial airliners spend more time in cruise speed and need more engine efficiency. High-performance fighter jets need very high acceleration and therefore have very high thrust requirements.

Weight

The weight of the aircraft is the common factor that links all aspects of aircraft design such as aerodynamics, structure, and propulsion together. An aircraft's weight is derived from various factors such as empty weight, payload, useful load, etc. The various weights are used to then calculate the center of mass of the entire aircraft. The center of mass must fit within the established limits set by the manufacturer.

Structure

The aircraft structure focuses not only on strength, stiffness, durability (fatigue), fracture toughness, stability, but also on fail-safety, corrosion resistance, maintainability and ease of manufacturing. The structure must be able to withstand the stresses caused by cabin pressurization, if fitted, turbulence and engine or rotor vibrations.

Design Process and Simulation

The design of any aircraft starts out in three phases:

Conceptual Design

The first design step, involves sketching a variety of possible aircraft configurations that meet the required design specifications. By drawing a set of configurations, designers seek to reach the design configuration that satisfactorily meets all requirements as well as go hand in hand with factors such as aerodynamics, propulsion, flight performance, structural and control systems. This is called design optimization. Fundamental aspects such as fuselage shape, wing configuration and location, engine size and type are all determined at this stage. Constraints to design like those mentioned above are all taken into account at this stage as well. The final product is a conceptual layout of the aircraft configuration on paper or computer screen, to be reviewed by engineers and other designers.

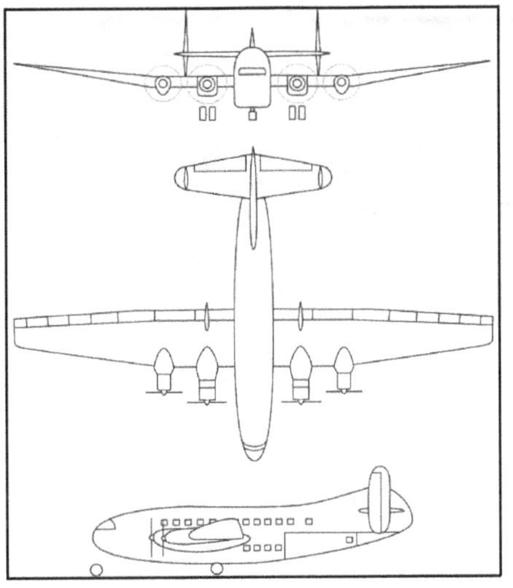

Conceptual design of a Breguet 673.

Preliminary Design Phase

The design configuration arrived at in the conceptual design phase is then tweaked and remodeled to fit into the design parameters. In this phase, wind tunnel testing and computational fluid dynamic calculations of the flow field around the aircraft are done. Major structural and control analysis is also carried out in this phase. Aerodynamic flaws and structural instabilities if any are corrected and the final design is drawn and finalized. Then after the finalization of the design lies the key decision with the manufacturer or individual designing it whether to actually go ahead with the production of the aircraft. At this point several designs, though perfectly capable of flight and performance, might have been opted out of production due to their being economically nonviable.

Detail Design Phase

This phase simply deals with the fabrication aspect of the aircraft to be manufactured. It determines the number, design and location of ribs, spars, sections and other structural elements. All aerodynamic, structural, propulsion, control and performance aspects have already been covered in the preliminary design phase and only the manufacturing remains. Flight simulators for aircraft are also developed at this stage.

Gull Wing

The gull wing is an aircraft wing configuration with a prominent bend in the wing inner section towards the wing root. Its name is derived from the seabirds which it resembles. It has been incorporated in aircraft for many reasons.

DFS Habicht glider showing gull wing profile.

Laughing gull showing the wing shape emulated in gull wing aircraft.

Sailplanes

The gull wing was first seen on a glider when the Weltensegler flew in 1921. Its wings were externally braced and featured swept-back wingtips. After the aircraft broke up, killing its pilot, the design feature stayed out of popular use. The gull wing made a resurgence in 1930 with Alexander Lippisch's record-breaking *Fafnir*. Lippisch used the configuration for its increased wingtip clearance and the ill-founded belief it improved stability in turns. However, studies have shown that normal gull wing configurations have significantly less severe and more easily recoverable stalls. Inverted gull wings show the opposite stall behaviour, but both normal and inverted gull wings impede lift/drag ratio and climb performance. The true success of the Fafnir's gull wing lay primarily in its aesthetic value; the gull wing would be a staple of the high-performance sailplanes of the time, until the 1950s.

Notable Gull Wing Sailplanes:

- Bowlus Senior Albatross
- DFS Habicht
- DFS Kranich
- DFS Reiher

- Göppingen Gö 3 *Minimoa*
- Lawrence Tech IV "Yankee Doodle"
- Lippisch Fafnir
- Ross RS-1 Zanonia
- Schweyer Rhönsperber
- Slingsby Kite
- Weltensegler

Seaplanes

Beriev Be-12 seaplane with gull wing profile.

The gull wing design found its way into seaplanes by the early 1930s. As engine power increased, so did the need for large propellers that could effectively convert power to thrust. The gull wing allowed designers to ensure adequate propeller tip clearance over the water by placing the engines on the highest point of the wing. The alternative was placing the engine on a pylon. Possibly the first flying boat to utilize the gull wing configuration was the Short Knuckleduster, which flew in 1933. The Dornier Do 26, a high-speed airliner and transport platform, of which 6 aircraft were built, flew in 1938. The configuration was also used on the US Navy's PBM Mariner and P5M Marlin maritime patrol aircraft. The emergence of long range, land-based jets in the 1950s and the subsequent demise of the seaplane prevented widespread use of the gull wing, although it was still used in some post-war designs, like Beriev Be-12 *Chaika* (the name means 'the gull' in Russian).

Examples:

- Beriev Be-6
- Dornier Do 26
- Martin P5M Marlin

- Piaggio P.136
- Short Knuckleduster

Landplanes

The gull wing design found its way into landplanes in the late 1920s, with Polish inventor Zygmunt Pulawski designing the PZL P.1 in 1928. The arrangement he devised is occasionally known as the "Pulawski Wing" or the "Polish wing". The gull wing was used to improve visibility in a high wing arrangement, because such wing could be thinnest by the fuselage, and in theory should limit pilot's view no more than A-pillars of a windscreen in a car body. It was used in fighter aircraft like PZL P.11 and Polikarpov I-15.

Examples:

- PZL P.11
- Polikarpov I-153

Inverted Gull Wing

Junkers Ju 87 Stuka German ground-attack aircraft of WWII.

F4U Corsair landing on USS *Bunker Hill*.

Aichi B7A carrying torpedo.

The inverted gull wing was also developed in the 1930s and was chiefly used on single engine military aircraft with increasing powerful engines. Before contra-rotating propellers came into use, such powers required larger diameter propellers but clearance between the propeller tip and ground had to be maintained. Long landing gear legs are heavy, bulky, and weaker than their shorter counterparts. The Vought F4U Corsair, designed from the onset as a carrier-based fighter, not only had the largest propeller of any U.S. fighter, but was also expected to face rough landings aboard a pitching carrier deck. The inverted gull wing allowed the landing gear to be short and strong, and to retract straight back, improving internal wing space. An additional aerodynamic advantage was that the wing/fuselage connection is perpendicular and has inherently lower drag than any other connection.

Another reason for having an inverted gull wing is to permit clearance for a large external bomb load, as on the Junkers Ju-87 Stuka.

Aircraft Flight Control System

A conventional fixed-wing aircraft flight control system consists of flight control surfaces, the respective cockpit controls, connecting linkages, and the necessary operating mechanisms to control an aircraft's direction in flight. Aircraft engine controls are also considered as flight controls as they change speed.

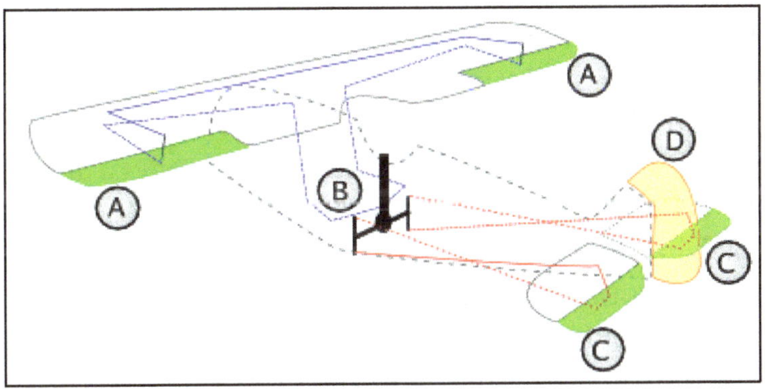

A typical aircraft's primary flight controls in motion.

The fundamentals of aircraft controls are explained in flight dynamics. This article centers on the operating mechanisms of the flight controls. The basic system in use on aircraft first appeared in a readily recognizable form as early as April 1908 on Louis Blériot's Blériot VIII pioneer-era monoplane design.

Cockpit Controls

Primary Controls

Generally, the primary cockpit flight controls are arranged as follows:

- a control yoke (also known as a control column), centre stick or side-stick (the latter two also colloquially known as a control or joystick), governs the aircraft's roll and pitch by moving the ailerons (or activating wing warping on some very early aircraft designs) when turned or deflected left and right, and moves the elevators when moved backwards or forwards.
- rudder pedals, or the earlier, pre-1919 "rudder bar", to control yaw, which move the rudder; left foot forward will move the rudder left for instance.
- throttle controls to control engine speed or thrust for powered aircraft.

The control yokes also vary greatly amongst aircraft. There are yokes where roll is controlled by rotating the yoke clockwise/counterclockwise (like steering a car) and pitch is controlled by tilting the control column towards you or away from you, but in others the pitch is controlled by sliding the yoke into and out of the instrument panel (like most Cessnas, such as the 152 and 172), and in some the roll is controlled by sliding the whole yoke to the left and right (like the Cessna 162). Centre sticks also vary between aircraft. Some are directly connected to the control surfaces using cables, others (fly-by-wire airplanes) have a computer in between which then controls the electrical actuators.

Blériot VIII at Issy-les-Moulineaux, the first flightworthy aircraft design to have the initial form of modern flight controls for the pilot.

Even when an aircraft uses variant flight control surfaces such as a V-tail ruddervator, flaperons, or elevons, to avoid pilot confusion the aircraft's flight control system will still be designed so that the stick or yoke controls pitch and roll conventionally, as will the rudder pedals for yaw. The basic pattern for modern flight controls was pioneered by French aviation figure Robert Esnault-Pelterie, with fellow French aviator Louis Blériot popularizing Esnault-Pelterie's control format initially on Louis' Blériot VIII monoplane in April 1908, and standardizing the format on the July 1909 Channel-crossing Blériot XI. Flight control has long been taught in such fashion for many decades, as popularized in ab initio instructional books such as the 1944 work Stick and Rudder.

In some aircraft, the control surfaces are not manipulated with a linkage. In ultralight aircraft and motorized hang gliders, for example, there is no mechanism at all. Instead, the pilot just grabs the lifting surface by hand (using a rigid frame that hangs from its underside) and moves it.

Secondary Controls

In addition to the primary flight controls for roll, pitch, and yaw, there are often secondary controls available to give the pilot finer control over flight or to ease the workload. The most commonly available control is a wheel or other device to control elevator trim, so that the pilot does not have to maintain constant backward or forward pressure to hold a specific pitch attitude (other types of trim, for rudder and ailerons, are common on larger aircraft but may also appear on smaller ones). Many aircraft have wing flaps, controlled by a switch or a mechanical lever or in some cases are fully automatic by computer control, which alter the shape of the wing for improved control at the slower speeds used for take-off and landing. Other secondary flight control systems may be available, including slats, spoilers, air brakes and variable-sweep wings.

Flight Control Systems

Mechanical

de Havilland Tiger Moth elevator and rudder cables.

Mechanical or manually operated flight control systems are the most basic method of controlling an aircraft. They were used in early aircraft and are currently used in small aircraft where the aerodynamic forces are not excessive. Very early aircraft, such as the Wright Flyer I, Blériot XI and Fokker Eindecker used a system of wing warping where no conventionally hinged control surfaces were used on the wing, and sometimes not even for pitch control as on the Wright Flyer I and original versions of the 1909 Etrich Taube, which only had a hinged/pivoting rudder in addition to the warping-operated pitch and roll controls. A manual flight control system uses a collection of mechanical parts such as pushrods, tension cables, pulleys, counterweights, and sometimes chains to transmit the forces applied to the cockpit controls directly to the control surfaces. Turnbuckles are often used to adjust control cable tension. The Cessna Skyhawk is a typical example of an aircraft that uses this type of system. Gust locks are often used on parked aircraft with mechanical systems to protect the control surfaces and linkages from damage from wind. Some aircraft have gust locks fitted as part of the control system.

Increases in the control surface area required by large aircraft or higher loads caused by high airspeeds in small aircraft lead to a large increase in the forces needed to move them, consequently complicated mechanical gearing arrangements were developed to extract maximum mechanical advantage in order to reduce the forces required from the pilots. This arrangement can be found on bigger or higher performance propeller aircraft such as the Fokker 50.

Some mechanical flight control systems use servo tabs that provide aerodynamic assistance. Servo tabs are small surfaces hinged to the control surfaces. The flight control mechanisms move these tabs, aerodynamic forces in turn move, or assist the movement of the control surfaces reducing the amount of mechanical forces needed. This arrangement was used in early piston-engined transport aircraft and in early jet transports. The Boeing 737 incorporates a system, whereby in the unlikely event of total hydraulic system failure, it automatically and seamlessly reverts to being controlled via servo-tab.

Hydro-mechanical

The complexity and weight of mechanical flight control systems increase considerably with the size and performance of the aircraft. Hydraulically powered control surfaces help to overcome these limitations. With hydraulic flight control systems, the aircraft's size and performance are limited by economics rather than a pilot's muscular strength. At first, only-partially boosted systems were used in which the pilot could still feel some of the aerodynamic loads on the control surfaces (feedback).

A hydro-mechanical flight control system has two parts:

- The *mechanical circuit*, which links the cockpit controls with the hydraulic circuits. Like the mechanical flight control system, it consists of rods, cables, pulleys, and sometimes chains.
- The *hydraulic circuit*, which has hydraulic pumps, reservoirs, filters, pipes, valves and actuators. The actuators are powered by the hydraulic pressure generated by the pumps in the hydraulic circuit. The actuators convert hydraulic pressure into control surface movements. The electro-hydraulic servo valves control the movement of the actuators.

The pilot's movement of a control causes the mechanical circuit to open the matching servo valve in the hydraulic circuit. The hydraulic circuit powers the actuators which then move the control surfaces. As the actuator moves, the servo valve is closed by a mechanical feedback linkage - one that stops movement of the control surface at the desired position.

This arrangement was found in the older-designed jet transports and in some high-performance aircraft. Examples include the Antonov An-225 and the Lockheed SR-71.

Artificial Feel Devices

With purely mechanical flight control systems, the aerodynamic forces on the control surfaces are transmitted through the mechanisms and are felt directly by the pilot, allowing tactile feedback of airspeed. With hydromechanical flight control systems, however, the load on the surfaces cannot be felt and there is a risk of overstressing the aircraft through excessive control surface movement. To overcome this problem, artificial feel systems can be used. For example, for the controls of the RAF's Avro Vulcan jet bomber and the RCAF's Avro Canada CF-105 Arrow supersonic interceptor (both 1950s-era designs), the required force feedback was achieved by a spring device. The fulcrum of this device was moved in proportion to the square of the air speed (for the elevators) to give increased resistance at higher speeds. For the controls of the American Vought F-8 Crusader and the LTV A-7 Corsair II warplanes, a 'bob-weight' was used in the pitch axis of the control stick, giving force feedback that was proportional to the airplane's normal acceleration.

Stick Shaker

A stick shaker is a device (available in some hydraulic aircraft) that is attached to the control column, which shakes the control column when the aircraft is about to stall. Also in some aircraft like the McDonnell Douglas DC-10 there is/was a back-up electrical power supply that the pilot can turn on to re-activate the stick shaker in case the hydraulic connection to the stick shaker is lost.

Fly-by-wire Control Systems

A fly-by-wire (FBW) system replaces manual flight control of an aircraft with an electronic interface. The movements of flight controls are converted to electronic signals transmitted by wires (hence the fly-by-wire term), and flight control computers determine how to move the actuators at each control surface to provide the expected response. Commands from the computers are also input without the pilot's knowledge to stabilize the aircraft and perform other tasks. Electronics for aircraft flight control systems are part of the field known as avionics.

Fly-by-optics, also known as *fly-by-light*, is a further development using fiber optic cables. This has an added advantage when sensitive electro-magnetic sensors will be operating aboard the aircraft.

Research

Several technology research and development efforts exist to integrate the functions of flight control systems such as ailerons, elevators, elevons, flaps, and flaperons into wings to perform the aerodynamic purpose with the advantages of less: mass, cost, drag, inertia (for faster, stronger control response), complexity (mechanically simpler, fewer moving parts or surfaces, less maintenance), and radar cross section for stealth. These may be used in many unmanned aerial vehicles (UAVs) and 6th generation fighter aircraft. Two promising approaches are flexible wings, and fluidics.

Flexible Wings

In flexible wings, much or all of a wing surface can change shape in flight to deflect air flow much like an ornithopter. Adaptive compliant wings are a military and commercial effort. The X-53 Active Aeroelastic Wing was a US Air Force, NASA, and Boeing effort.

Fluidics

In fluidics, forces in vehicles occur via circulation control, in which larger more complex mechanical parts are replaced by smaller simpler fluidic systems (slots which emit air flows) where larger forces in fluids are diverted by smaller jets or flows of fluid intermittently, to change the direction of vehicles. In this use, fluidics promises lower mass, costs (up to 50% less), and very low inertia and response times, and simplicity. This was demonstrated in the Demon UAV, which flew for the first time, in the UK, in September 2010.

Aviation

Aviation is the practical aspect or art of aeronautics, being the design, development, production, operation and use of aircraft, especially heavier than air aircraft. The word *aviation* was coined by French writer and former naval officer Gabriel La Landelle in 1863, from the verb *avier* (synonymous flying), itself derived from the Latin word *avis* ("bird") and the suffix *-ation*.

Early Beginnings

There are early legends of human flight such as the story of Icarus in Greek myth and Jamshid in Persian myth, and later, somewhat more credible claims of short-distance human flights appear, such as the flying automaton of Archytas of Tarentum (428–347 BC), the winged flights of Abbas Ibn Firnas (810–887), Eilmer of Malmesbury (11th century), and the hot-air Passarola of Bartholomeu Lourenço de Gusmão (1685–1724).

Lighter than Air

LZ 129 Hindenburg at Lakehurst Naval Air Station, 1936.

The modern age of aviation began with the first untethered human lighter-than-air flight on November 21, 1783, of a hot air balloon designed by the Montgolfier brothers. The practicality of balloons was limited because they could only travel downwind. It was immediately recognized that a steerable, or dirigible, balloon was required. Jean-Pierre Blanchard flew the first human-powered dirigible in 1784 and crossed the English Channel in one in 1785.

Rigid airships became the first aircraft to transport passengers and cargo over great distances. The best known aircraft of this type were manufactured by the German Zeppelin company.

The most successful Zeppelin was the Graf Zeppelin. It flew over one million miles, including an around-the-world flight in August 1929. However, the dominance of the Zeppelins over the airplanes of that period, which had a range of only a few hundred miles, was diminishing as airplane design advanced. The "Golden Age" of the airships ended on May 6, 1937 when the Hindenburg caught fire, killing 36 people. The cause of the Hindenburg accident was initially blamed on the use of hydrogen instead of helium as the lift gas. An internal investigation by the manufacturer revealed the coating used to protect the covering material over the frame was highly flammable and allowed static electricity to build up in the airship. Changes to the coating formulation reduced the risk of further Hindenburg type accidents. Although there have been periodic initiatives to revive their use, airships have seen only niche application since that time.

Heavier than Air

In 1799 Sir George Cayley set forth the concept of the modern airplane as a fixed-wing flying machine with separate systems for lift, propulsion, and control. Early dirigible developments included machine-powered propulsion (Henri Giffard, 1852), rigid frames (David Schwarz, 1896) and improved speed and maneuverability (Alberto Santos-Dumont, 1901).

First flight by the Wright Brothers, December 17, 1903.

There are many competing claims for the earliest powered, heavier-than-air flight. The first recorded powered flight was carried out by Clément Ader on October 9,

1890 when he reportedly made the first manned, powered, heavier-than-air flight of a significant distance (50 m (160 ft)) but insignificant altitude from level ground in his bat-winged, fully self-propelled fixed-wing aircraft, the Ader Éole. Seven years later, on 14 October 1897, Ader's Avion III was tested without success in front of two officials from the French War ministry. The report on the trials was not publicized until 1910, as they had been a military secret. In November 1906 Ader claimed to have made a successful flight on 14 October 1897, achieving an "uninterrupted flight" of around 300 metres (980 feet) on. Although widely believed at the time, these claims were later discredited.

However, the most widely accepted date is December 17, 1903 by the Wright brothers. The Wright brothers were the first to fly in a powered and controlled aircraft. Previous flights were gliders (control but no power) or free flight (power but no control), but the Wright brothers combined both, setting the new standard in aviation records. Following this, the widespread adoption of ailerons rather than wing warping made aircraft much easier to control, and only a decade later, at the start of World War I, heavier-than-air powered aircraft had become practical for reconnaissance, artillery spotting, and even attacks against ground positions.

Aircraft began to transport people and cargo as designs grew larger and more reliable. The Wright brothers took aloft the first passenger, Charles Furnas, one of their mechanics, on May 14, 1908.

During the 1920s and 1930s great progress was made in the field of aviation, including the first transatlantic flight of Alcock and Brown in 1919, Charles Lindbergh's solo transatlantic flight in 1927, and Charles Kingsford Smith's transpacific flight the following year. One of the most successful designs of this period was the Douglas DC-3, which became the first airliner to be profitable carrying passengers exclusively, starting the modern era of passenger airline service. By the beginning of World War II, many towns and cities had built airports, and there were numerous qualified pilots available. The war brought many innovations to aviation, including the first jet aircraft and the first liquid-fueled rockets.

NASA's Helios researches solar powered flight.

After World War II, especially in North America, there was a boom in general aviation, both private and commercial, as thousands of pilots were released from military service and many inexpensive war-surplus transport and training aircraft became available. Manufacturers such as Cessna, Piper, and Beechcraft expanded production to provide light aircraft for the new middle-class market.

By the 1950s, the development of civil jets grew, beginning with the de Havilland Comet, though the first widely used passenger jet was the Boeing 707, because it was much more economical than other aircraft at that time. At the same time, turboprop propulsion began to appear for smaller commuter planes, making it possible to serve small-volume routes in a much wider range of weather conditions.

Since the 1960s composite material airframes and quieter, more efficient engines have become available, and Concorde provided supersonic passenger service for more than two decades, but the most important lasting innovations have taken place in instrumentation and control. The arrival of solid-state electronics, the Global Positioning System, satellite communications, and increasingly small and powerful computers and LED displays, have dramatically changed the cockpits of airliners and, increasingly, of smaller aircraft as well. Pilots can navigate much more accurately and view terrain, obstructions, and other nearby aircraft on a map or through synthetic vision, even at night or in low visibility.

On June 21, 2004, SpaceShipOne became the first privately funded aircraft to make a spaceflight, opening the possibility of an aviation market capable of leaving the Earth's atmosphere. Meanwhile, flying prototypes of aircraft powered by alternative fuels, such as ethanol, electricity, and even solar energy, are becoming more common.

Operations of Aircraft

Civil Aviation

Diamond Aircraft Industries Diamond D-Jet (very light jet).

Civil aviation includes all non-military flying, both general aviation and scheduled air transport.

Air Transport

Northwest Airlines Airbus A330-323X.

There are five major manufacturers of civil transport aircraft (in alphabetical order):

- Airbus, based in Europe.
- Boeing, based in the United States.
- Bombardier, based in Canada.
- Embraer, based in Brazil.
- United Aircraft Corporation, based in Russia.

Boeing, Airbus, Ilyushin and Tupolev concentrate on wide-body and narrow-body jet airliners, while Bombardier, Embraer and Sukhoi concentrate on regional airliners. Large networks of specialized parts suppliers from around the world support these manufacturers, who sometimes provide only the initial design and final assembly in their own plants. The Chinese ACAC consortium will also soon enter the civil transport market with its Comac ARJ21 regional jet.

Until the 1970s, most major airlines were flag carriers, sponsored by their governments and heavily protected from competition. Since then, open skies agreements have resulted in increased competition and choice for consumers, coupled with falling prices for airlines. The combination of high fuel prices, low fares, high salaries, and crises such as the September 11, 2001 attacks and the SARS epidemic have driven many older airlines to government-bailouts, bankruptcy or mergers. At the same time, low-cost carriers such as Ryanair, Southwest and Westjet have flourished.

General Aviation

General aviation includes all non-scheduled civil flying, both private and commercial. General aviation may include business flights, air charter, private aviation, flight training, ballooning, parachuting, gliding, hang gliding, aerial photography, foot-launched powered hang gliders, air ambulance, crop dusting, charter flights, traffic reporting, police air patrols and forest fire fighting.

1947 Cessna 120.

A weight-shift ultralight aircraft, the Air Creation Tanarg.

Each country regulates aviation differently, but general aviation usually falls under different regulations depending on whether it is private or commercial and on the type of equipment involved.

Many small aircraft manufacturers serve the general aviation market, with a focus on private aviation and flight training.

The most important recent developments for small aircraft (which form the bulk of the GA fleet) have been the introduction of advanced avionics (including GPS) that were formerly found only in large airliners, and the introduction of composite materials to make small aircraft lighter and faster. Ultralight and homebuilt aircraft have also become increasingly popular for recreational use, since in most countries that allow private aviation, they are much less expensive and less heavily regulated than certified aircraft.

The largest aircraft to be built, to date, is the Antonov An-225. This aircraft comes from the Ukraine, and it was built back in the 1980s. This aircraft includes 6 engines, mounted on the wing. Its wingspan is 88 metres (289 feet) and it is 84 metres (276 feet) long. This aircraft holds the world payload record, after it transported 428,834 pounds worth of goods. Weighing in at 1.4 million pounds, it is also the heaviest aircraft to be built.

Military Aviation

Simple balloons were used as surveillance aircraft as early as the 18th century. Over the years, military aircraft have been built to meet ever increasing capability requirements.

Manufacturers of military aircraft compete for contracts to supply their government's arsenal. Aircraft are selected based on factors like cost, performance, and the speed of production.

The Lockheed SR-71 remains unsurpassed in many areas of performance.

Types of Military Aviation

- Fighter aircraft's primary function is to destroy other aircraft. (e.g. Sopwith Camel, A6M Zero, F-15, MiG-29, Su-27, and F-22).
- Ground attack aircraft are used against tactical earth-bound targets. (e.g. Junkers Stuka, A-10, Il-2, J-22 Orao, AH-64 and Su-25).
- Bombers are generally used against more strategic targets, such as factories and oil fields. (e.g. Zeppelin, Tu-95, Mirage IV, and B-52).
- Transport aircraft are used to transport hardware and personnel. (e.g. C-17 Globemaster III, C-130 Hercules and Mil Mi-26).
- Surveillance and reconnaissance aircraft obtain information about enemy forces. (e.g. Rumpler Taube, Mosquito, U-2, OH-58 and MiG-25R).
- Unmanned aerial vehicles (UAVs) are used primarily as reconnaissance fixed-wing aircraft, though many also carry payloads. Cargo aircraft are in development. (e.g. RQ-7B Shadow, MQ-8 Fire Scout, and MQ-1C Gray Eagle).
- Missiles deliver warheads, normally explosives, but also things like leaflets.

Air Safety

Aviation Accidents and Incidents

An *aviation accident* is defined by the Convention on International Civil Aviation Annex 13 as an occurrence associated with the operation of an aircraft which takes place between the time any person boards the aircraft with the intention of flight until such time as all such persons have disembarked, in which a person is fatally or seriously injured, the aircraft sustains damage or structural failure or the aircraft is missing or is completely inaccessible.

A USAF Thunderbird pilot ejecting from his F-16 aircraft at an airshow in 2003.

The first fatal aviation accident occurred in a Wright Model A aircraft at Fort Myer, Virginia, USA, on September 17, 1908, resulting in injury to the pilot, Orville Wright, and death of the passenger, Signal Corps Lieutenant Thomas Selfridge.

An *aviation incident* is defined as an occurrence, other than an accident, associated with the operation of an aircraft that affects or could affect the safety of operations.

An accident in which the damage to the aircraft is such that it must be written off, or in which the plane is destroyed, is called a *hull loss accident*.

Air Traffic Control

Air traffic control towers at Amsterdam Airport.

Air traffic control (ATC) involves communication with aircraft to help maintain separation — that is, they ensure that aircraft are sufficiently far enough apart horizontally or vertically for no risk of collision. Controllers may co-ordinate position reports provided by pilots, or in high traffic areas (such as the United States) they may use radar to see aircraft positions.

There are generally four different types of ATC:

- center controllers, who control aircraft en route between airports.
- control towers (including tower, ground control, clearance delivery, and other services), which control aircraft within a small distance (typically 10–15 km horizontal, and 1,000 m vertical) of an airport.
- oceanic controllers, who control aircraft over international waters between continents, generally without radar service.
- terminal controllers, who control aircraft in a wider area (typically 50–80 km) around busy airports.

ATC is especially important for aircraft flying under instrument flight rules (IFR), where they may be in weather conditions that do not allow the pilots to see other aircraft. However, in very high-traffic areas, especially near major airports, aircraft flying under visual flight rules (VFR) are also required to follow instructions from ATC.

In addition to separation from other aircraft, ATC may provide weather advisories, terrain separation, navigation assistance, and other services to pilots, depending on their workload.

ATC do not control all flights. The majority of VFR flights in North America are not required to talk to ATC (unless they are passing through a busy terminal area or using a major airport), and in many areas, such as northern Canada and low altitude in northern Scotland, Air traffic control services are not available even for IFR flights at lower altitudes.

Environmental Impact

Water vapor contrails left by high-altitude jet airliners. These may contribute to cirrus cloud formation.

Like all activities involving combustion, operating powered aircraft (from airliners to hot air balloons) release soot and other pollutants into the atmosphere. Greenhouse gases such as carbon dioxide (CO_2) are also produced. In addition, there are environmental impacts specific to aviation:

- Aircraft operating at high altitudes near the tropopause (mainly large jet airliners) emit aerosols and leave contrails, both of which can increase cirrus cloud formation — cloud cover may have increased by up to 0.2% since the birth of aviation.
- Aircraft operating at high altitudes near the tropopause can also release chemicals that interact with greenhouse gases at those altitudes, particularly nitrogen compounds, which interact with ozone, increasing ozone concentrations.
- Most light piston aircraft burn avgas, which contains tetraethyllead (TEL). Some lower-compression piston engines can operate on unleaded mogas, and turbine engines and diesel engines — neither of which requires lead — are appearing on some newer light aircraft.

Another environmental impact of aviation is noise pollution, mainly caused by aircraft taking off and landing.

Aeronautics

Space Shuttle *Atlantis* on a Shuttle Carrier Aircraft.

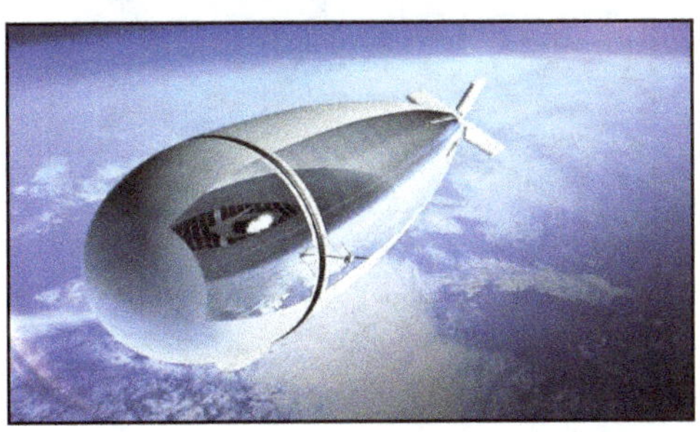

Lighter than air geostationary airship telecommunications satellite.

Aeronautics (from the ancient Greek words □ήρ *āēr*, which means "air", and ναυτική *nautikē* which means "navigation", i.e. "navigation into the air") is the science or art involved with the study, design, and manufacturing of air flight capable machines, and the techniques of operating aircraft and rockets within the atmosphere. The British Royal Aeronautical Society identifies the aspects of "aeronautical Art, Science and Engineering" and "the profession of Aeronautics (which expression includes Astronautics)."

While the term—literally meaning "sailing the air"—originally referred solely to the science of *operating* the aircraft, it has since been expanded to include technology, business and other aspects related to aircraft. The term "aviation" is sometimes used interchangeably with aeronautics, although "aeronautics" includes lighter-than-air craft such as airships, and includes ballistic vehicles while "aviation" technically does not.

A significant part of aeronautical science is a branch of dynamics called aerodynamics, which deals with the motion of air and the way that it interacts with objects in motion, such as an aircraft.

History

Early Ideas

Attempts to fly without any real aeronautical understanding have been made from the earliest times, typically by constructing wings and jumping from a tower with crippling or lethal results.

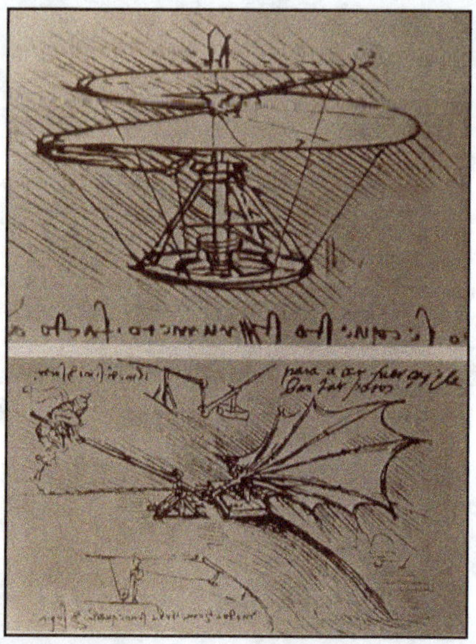

Designs for flying machines by Leonardo da Vinci, circa 1490.

Wiser investigators sought to gain some rational understanding through the study of bird flight. An early example appears in ancient Egyptian texts. Later medieval Islamic scientists also made such studies. The founders of modern aeronautics, Leonardo da Vinci in the Renaissance and Cayley in 1799, both began their investigations with studies of bird flight.

Man-carrying kites are believed to have been used extensively in ancient China. In 1282 the European explorer Marco Polo described the Chinese techniques then current. The Chinese also constructed small hot air balloons, or lanterns, and rotary-wing toys.

An early European to provide any scientific discussion of flight was Roger Bacon, who described principles of operation for the lighter-than-air balloon and the flapping-wing ornithopter, which he envisaged would be constructed in the future. The lifting medium for his balloon would be an "aether" whose composition he did not know.

In the late fifteenth century, Leonardo da Vinci followed up his study of birds with designs for some of the earliest flying machines, including the flapping-wing ornithopter and the rotating-wing helicopter. Although his designs were rational, they were not based on particularly good science. Many of his designs, such as a four-person screw-type helicopter, have severe flaws. He did at least understand that "An object offers as much resistance to the air as the air does to the object." (Newton would not publish the Third law of motion until 1687.) His analysis led to the realisation that manpower alone was not sufficient for sustained flight, and his later designs included a mechanical power source such as a spring. Da Vinci's work was lost after his death and did not reappear until it had been overtaken by the work of George Cayley.

Balloon Flight

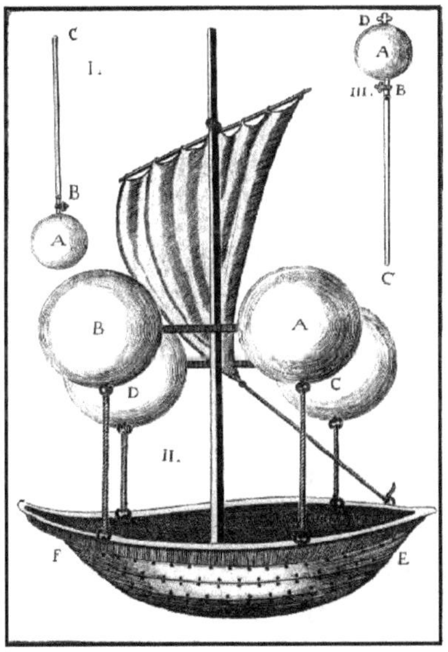

Francesco Lana de Terzi's flying boat concept c.1670.

The modern era of lighter-than-air flight began early in the 17th century with Galileo's experiments in which he showed that air has weight. Around 1650 Cyrano de Bergerac wrote some fantasy novels in which he described the principle of ascent using a substance (dew) he supposed to be lighter than air, and descending by releasing a controlled amount of the substance. Francesco Lana de Terzi measured the pressure of air at sea level and in 1670 proposed the first scientifically credible lifting medium in the form of hollow metal spheres from which all the air had been pumped out. These would be lighter than the displaced air and able to lift an airship. His proposed methods of controlling height are still in use today; by carrying ballast which may be dropped overboard to gain height, and by venting the lifting containers to lose height. In practice de Terzi's spheres would have collapsed under air pressure, and further developments had to wait for more practicable lifting gases.

From the mid-18th century the Montgolfier brothers in France began experimenting with balloons. Their balloons were made of paper, and early experiments using steam as the lifting gas were short-lived due to its effect on the paper as it condensed. Mistaking smoke for a kind of steam, they began filling their balloons with hot smoky air which they called "electric smoke" and, despite not fully understanding the principles at work, made some successful launches and in 1783 were invited to give a demonstration to the French Academie des Sciences.

Meanwhile, the discovery of hydrogen led Joseph Black in c. 1780 to propose its use as a lifting gas, though practical demonstration awaited a gastight balloon material. On hearing of the Montgolfier Brothers' invitation, the French Academy member Jacques Charles offered a similar demonstration of a hydrogen balloon. Charles and two craftsmen, the Robert brothers, developed a gastight material of rubberised silk for the envelope. The hydrogen gas was to be generated by chemical reaction during the filling process.

The Montgolfier designs had several shortcomings, not least the need for dry weather and a tendency for sparks from the fire to set light to the paper balloon. The manned design had a gallery around the base of the balloon rather than the hanging basket of the first, unmanned design, which brought the paper closer to the fire. On their free flight, De Rozier and d'Arlandes took buckets of water and sponges to douse these fires as they arose. On the other hand, the manned design of Charles was essentially modern. As a result of these exploits, the hot-air ballon became known as the *Montgolfière* type and the hydrogen balloon the *Charlière*.

Charles and the Robert brothers' next balloon, *La Caroline*, was a Charlière that followed Jean Baptiste Meusnier's proposals for an elongated dirigible balloon, and was notable for having an outer envelope with the gas contained in a second, inner ballonet. On 19 September 1784, it completed the first flight of over 100 km, between Paris and Beuvry, despite the man-powered propulsive devices proving useless.

In an attempt the next year to provide both endurance and controllability, de Rozier developed a balloon having both hot air and hydrogen gas bags, a design which was soon named after him as the *Rozière*. The principle was to use the hydrogen section for constant lift and to navigate vertically by heating and allowing to cool the hot air section, in order to catch the most favourable wind at whatever altitude it was blowing. The balloon envelope was made of goldbeaters skin. The first flight ended in disaster and the approach has seldom been used since.

Cayley and the Foundation of Modern Aeronautics

Sir George Cayley (1773-1857) is widely acknowledged as the founder of modern aeronautics. He was first called the "father of the aeroplane" in 1846 and Henson called him the "father of aerial navigation." He was the first true scientific aerial investigator to publish his work, which included for the first time the underlying principles and forces of flight.

In 1809 he began the publication of a landmark three-part treatise titled "On Aerial Navigation" (1809–1810). In it he wrote the first scientific statement of the problem, "The whole problem is confined within these limits, viz. to make a surface support a given weight by the application of power to the resistance of air." He identified the four vector forces that influence an aircraft: *thrust*, *lift*, *drag* and *weight* and distinguished stability and control in his designs.

He developed the modern conventional form of the fixed-wing aeroplane having a stabilising tail with both horizontal and vertical surfaces, flying gliders both unmanned and manned.

He introduced the use of the whirling arm test rig to investigate the aerodynamics of flight, using it to discover the benefits of the curved or cambered aerofoil over the flat wing he had used for his first glider. He also identified and described the importance of dihedral, diagonal bracing and drag reduction, and contributed to the understanding and design of ornithopters and parachutes.

Another significant invention was the tension-spoked wheel, which he devised in order to create a light, strong wheel for aircraft undercarriage.

The 19th Century

During the 19th century Cayley's ideas were refined, proved and expanded on. Important investigators included Otto Lilienthal and Horatio Phillips.

Branches

The Eurofighter Typhoon.

Antonov An-225 *Mriya*, the largest aeroplane ever built.

Aeronautics may be divided into three main branches comprising Aviation, Aeronautical science and Aeronautical engineering.

Aviation

Aviation is the art or practice of aeronautics. Historically aviation meant only heavier-than-air flight, but nowadays it includes flying in balloons and airships.

Aeronautical Science

Aeronautical science covers the practical theory of aeronautics and aviation, including operations, navigation, air safety and human factors.

A candidate pilot is likely to study for a qualification in aeronautical science.

Aeronautical Engineering

Aeronautical engineering covers the design and construction of aircraft, including how they are powered, how they are used and how they are controlled for safe operation.

A major part of aeronautical engineering is aerodynamics, the science of passing through the air.

With the increasing activity in spaceflight, nowadays aeronautics and astronautics are often combined as aerospace engineering.

Aerodynamics

The science of aerodynamics deals with the motion of air and the way that it interacts with objects in motion, such as an aircraft.

The study of aerodynamics falls broadly into three areas:

Incompressible flow occurs where the air simply moves to avoid objects, typically at subsonic speeds below that of sound (Mach 1).

Compressible flow occurs where shock waves appear at points where the air becomes compressed, typically at speeds above Mach 1.

Transonic flow occurs in the intermediate speed range around Mach 1, where the airflow over an object may be locally subsonic at one point and locally supersonic at another.

Rocketry

A rocket or rocket vehicle is a missile, spacecraft, aircraft or other vehicle which obtains thrust from a rocket engine. In all rockets, the exhaust is formed entirely from propellants carried within the rocket before use. Rocket engines work by action and reaction. Rocket engines push rockets forwards simply by throwing their exhaust backwards extremely fast.

Rockets for military and recreational uses date back to at least 13th century China. Significant scientific, interplanetary and industrial use did not occur until the 20th century, when rocketry was the enabling technology of the Space Age, including setting foot on the moon.

Rockets are used for fireworks, weaponry, ejection seats, launch vehicles for artificial satellites, human spaceflight and exploration of other planets. While comparatively inefficient for low speed use, they are very lightweight and powerful, capable of generating large accelerations and of attaining extremely high speeds with reasonable efficiency.

Chemical rockets are the most common type of rocket and they typically create their exhaust by the combustion of rocket propellant. Chemical rockets store a large amount of energy in an easily released form, and can be very dangerous. However, careful design, testing, construction and use minimizes risks.

References

- Cayley, George. "On Aerial Navigation" Part 1, Part 2, Part 3 Nicholson's Journal of Natural Philosophy, 1809–1810. (Via NASA). Raw text. Retrieved: 30 May 2020
- Sutton, George (2001). "1". Rocket Propulsion Elements (7th ed.). Chichester: John Wiley & Sons. ISBN 978-0-471-32642-7

4

Significant Concepts of Aircraft Aerodynamics

The study of the motion of air especially when it is affected by a solid object is termed as aerodynamics. A few areas of study, under it are lift force, aerodynamic drag, lift-induced drag, center of pressure, boundary layer and compressible flow. All these diverse concepts related to aerodynamics have been carefully analyzed in this chapter.

Aerodynamics is the study of how gases interact with moving bodies. Because the gas that we encounter most is air, aerodynamics is primarily concerned with the forces of drag and lift, which are caused by air passing over and around solid bodies. Engineers apply the principles of aerodynamics to the designs of many different things, including buildings, bridges and even soccer balls; however, of primary concern is the aerodynamics of aircraft and automobiles.

Aerodynamics comes into play in the study of flight and the science of building and operating an aircraft, which is called aeronautics. Aeronautical engineers use the fundamentals of aerodynamics to design aircraft that fly through the Earth's atmosphere.

Aerodynamic Drag

The most significant aerodynamic force that applies to nearly everything that moves through the air is drag. Drag is the force that opposes an aircraft's motion through the air, according to NASA. Drag is generated in the direction the air is moving when it encounters a solid object. In most cases, such as in automobiles and aircraft, drag is undesirable because it takes power to overcome it. There are, however, some cases when drag is beneficial, such as with parachutes, for example.

To describe the amount of drag on an object, we use a value called the drag coefficient (c_d). This number depends not only on the shape of the object but also on other factors, such as its speed and surface roughness, the density of the air and whether the flow is laminar (smooth) or turbulent. Forces that affect drag include the air pressure against the face of the object, the friction along the sides of the object and the relatively negative pressure, or suction, on the back of the object. For example, c_d for a flat plate moving face-on through the air is about 1.3, a face-on cube is about 1, a sphere is about 0.5 and a teardrop shape is about 0.05. The drag coefficient for modern automobiles is 0.25 to 0.35, and for aircraft it is 0.01 to 0.03. Calculating c_d can be complicated. For this reason, it is usually determined by computer simulations or wind tunnel experiments.

Fundamental Concepts

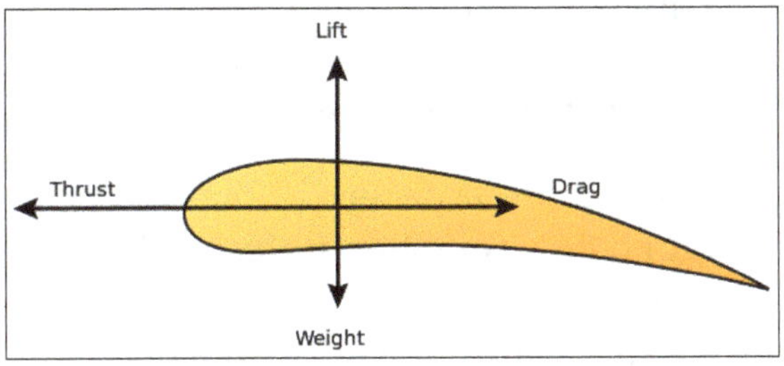

Forces of flight on an airfoil.

Understanding the motion of air around an object (often called a flow field) enables the calculation of forces and moments acting on the object. In many aerodynamics problems, the forces of interest are the fundamental forces of flight: lift, drag, thrust, and weight. Of these, lift and drag are aerodynamic forces, i.e. forces due to air flow over a solid body. Calculation of these quantities is often founded upon the assumption that the flow field behaves as a continuum. Continuum flow fields are characterized by properties such as flow velocity, pressure, density, and temperature, which may be functions of position and time. These properties may be directly or indirectly measured in aerodynamics experiments or calculated starting with the equations for conservation of mass, momentum, and energy in air flows. Density, flow velocity, and an additional property, viscosity, are used to classify flow fields.

Flow Classification

Flow velocity is used to classify flows according to speed regime. Subsonic flows are flow fields in which the air speed field is always below the local speed of sound. Transonic flows include both regions of subsonic flow and regions in which the local flow speed is greater than the local speed of sound. Supersonic flows are defined to be flows in which the flow speed is greater than the speed of sound everywhere. A fourth classification, hypersonic flow, refers to flows where the flow speed is much greater than the speed of sound. Aerodynamicists disagree on the precise definition of hypersonic flow.

Compressible flow accounts for varying density within the flow. Subsonic flows are often idealized as incompressible, i.e. the density is assumed to be constant. Transonic and supersonic flows are compressible, and calculations that neglect the changes of density in these flow fields will yield inaccurate results.

Viscosity is associated with the frictional forces in a flow. In some flow fields, viscous effects are very small, and approximate solutions may safely neglect viscous effects. These approximations are called inviscid flows. Flows for which viscosity is not neglected are called viscous flows. Finally, aerodynamic problems may also be classified by the flow environment. External aerodynamics is the study of flow around solid objects of various shapes (e.g. around an airplane wing), while internal aerodynamics is the study of flow through passages inside solid objects (e.g. through a jet engine).

Continuum Assumption

Unlike liquids and solids, gases are composed of discrete molecules which occupy only a small fraction of the volume filled by the gas. On a molecular level, flow fields are made up of the collisions of many individual of gas molecules between themselves and with solid surfaces. However, in most aerodynamics applications, the discrete molecular nature of gases is ignored, and the flow field is assumed to behave as a continuum. This assumption allows fluid properties such as density and flow velocity to be defined everywhere within the flow.

The validity of the continuum assumption is dependent on the density of the gas and the application in question. For the continuum assumption to be valid, the mean free path length must be much smaller than the length scale of the application in question. For example, many aerodynamics applications deal with aircraft flying in atmospheric conditions, where the mean free path length is on the order of micrometers and where the body is orders of magnitude larger. In these cases, the length scale of the aircraft ranges from a few meters to a few tens of meters, which is much larger than the mean free path length. For such applications, the continuum assumption is reasonable. The continuum assumption is less valid for extremely low-density flows, such as those encountered by vehicles at very high altitudes (e.g. 300,000 ft/90 km) or satellites in Low Earth orbit. In those cases, statistical mechanics is a more accurate method of solving the problem than is continuum aerodynamics. The Knudsen number can be used to guide the choice between statistical mechanics and the continuous formulation of aerodynamics.

Conservation Laws

The assumption of a fluid continuum allows problems in aerodynamics to be solved using fluid dynamics conservation laws. Three conservation principles are used:

1. Conservation of mass: In fluid dynamics, the mathematical formulation of this principle is known as the mass continuity equation, which requires that mass is neither created nor destroyed within a flow of interest.
2. Conservation of momentum: In fluid dynamics, the mathematical formulation of this principle can be considered an application of Newton's Second Law. Momentum within a flow is only changed by the work performed on the system by external forces, which may include both surface forces, such as viscous (frictional) forces, and body forces, such as weight. The momentum conservation principle may be expressed as either a vector equation or separated into a set of three scalar equations (x,y,z components). In its most complete form, the momentum conservation equations are known as the Navier-Stokes equations. The Navier-Stokes equations have no known analytical solution and are solved in modern aerodynamics using computational techniques. Because of the computational cost of solving these complex equations, simplified expressions of momentum conservation may be appropriate for specific applications. The Euler equations are a set of momentum conservation equations which neglect viscous forces and may be used in cases where the effect of viscous forces is expected to be small. Additionally, Bernoulli's equation is a solution to the momentum conservation equation of an inviscid flow that neglects gravity.
3. Conservation of energy: The energy conservation equation states that energy is neither created nor destroyed within a flow, and that any addition or subtraction of energy to a volume in the flow is caused by the fluid flow, by heat transfer, or by work into and out of the region of interest.

The ideal gas law or another such equation of state is often used in conjunction with these equations to form a determined system that allows the solution for the unknown variables.

Branches of Aerodynamics

Aerodynamic problems are classified by the flow environment or properties of the flow, including flow speed, compressibility, and viscosity. *External* aerodynamics is the study of flow around solid objects of various shapes. Evaluating the lift and drag on an airplane or the shock waves that form in front of the nose of a rocket are examples of external aerodynamics. *Internal* aerodynamics is the study of flow through passages in solid objects. For instance, internal aerodynamics encompasses the study of the airflow through a jet engine or through an air conditioning pipe.

Aerodynamic problems can also be classified according to whether the flow speed is below, near or above the speed of sound. A problem is called subsonic if all the speeds in the problem are less than the speed of sound, transonic if speeds both below and above the speed of sound are present (normally when the characteristic speed is approximately the speed of sound), supersonic when the characteristic flow speed is greater than the speed of sound, and hypersonic when the flow speed is much greater than the speed of sound. Aerodynamicists disagree over the precise definition of hypersonic flow; a rough definition considers flows with Mach numbers above 5 to be hypersonic.

The influence of viscosity on the flow dictates a third classification. Some problems may encounter only very small viscous effects, in which case viscosity can be considered to be negligible. The approximations to these problems are called inviscid flows. Flows for which viscosity cannot be neglected are called viscous flows.

Incompressible Aerodynamics

An incompressible flow is a flow in which density is constant in both time and space. Although all real fluids are compressible, a flow is often approximated as incompressible if the effect of the density changes cause only small changes to the calculated results. This is more likely to be true when the flow speeds are significantly lower than the speed of sound. Effects of compressibility are more significant at speeds close to or above the speed of sound. The Mach number is used to evaluate whether the incompressibility can be assumed, otherwise the effects of compressibility must be included.

Subsonic Flow

Subsonic (or low-speed) aerodynamics describes fluid motion in flows which are much lower than the speed of sound everywhere in the flow. There are several branches of subsonic flow but one special case arises when the flow is inviscid, incompressible and irrotational. This case is called potential flow and allows the differential equations that describe the flow to be a simplified version of the equations of fluid dynamics, thus making available to the aerodynamicist a range of quick and easy solutions.

In solving a subsonic problem, one decision to be made by the aerodynamicist is whether to incorporate the effects of compressibility. Compressibility is a description of the amount of change of density in the flow. When the effects of compressibility on the solution are small, the assumption that density is constant may be made. The problem is then an incompressible low-speed aerodynamics problem. When the density is allowed to vary, the flow is called compressible. In air, compressibility effects are usually ignored when the Mach number in the flow does not exceed 0.3 (about 335 feet (102 m) per second or 228 miles (366 km) per hour at 60 °F (16 °C)). Above Mach 0.3, the problem flow should be described using compressible aerodynamics.

Compressible Aerodynamics

According to the theory of aerodynamics, a flow is considered to be compressible if the density changes along a streamline. This means that – unlike incompressible flow – changes in density are considered. In general, this is the case where the Mach number in part or all of the flow exceeds 0.3. The Mach 0.3 value is rather arbitrary, but it is used because gas flows with a Mach number below that value demonstrate changes in density of less than 5%. Furthermore, that maximum 5% density change occurs at the stagnation point (the point on the object where flow speed is zero), while the density changes around the rest of the object will be significantly lower. Transonic, supersonic, and hypersonic flows are all compressible flows.

Transonic Flow

The term Transonic refers to a range of flow velocities just below and above the local speed of sound (generally taken as Mach 0.8–1.2). It is defined as the range of speeds between the critical Mach number, when some parts of the airflow over an aircraft become supersonic, and a higher speed, typically near Mach 1.2, when all of the airflow is supersonic. Between these speeds, some of the airflow is supersonic, while some of the airflow is not supersonic.

Supersonic Flow

Supersonic aerodynamic problems are those involving flow speeds greater than the speed of sound. Calculating the lift on the Concorde during cruise can be an example of a supersonic aerodynamic problem.

Supersonic flow behaves very differently from subsonic flow. Fluids react to differences in pressure; pressure changes are how a fluid is "told" to respond to its environment. Therefore, since sound is in fact an infinitesimal pressure difference propagating through a fluid, the speed of sound in that fluid can be considered the fastest speed that "information" can travel in the flow. This difference most obviously manifests itself in the case of a fluid striking an object. In front of that object, the fluid builds up a stagnation pressure as impact with the object brings the moving fluid to rest. In fluid traveling at subsonic speed, this pressure disturbance can propagate upstream, changing the flow pattern ahead of the object and giving the impression that the fluid "knows" the object is there by seemingly adjusting its movement and is flowing around it. In a supersonic flow however, the pressure disturbance cannot propagate upstream. Thus, when the fluid finally reaches the object it strikes it and the fluid is forced to change its properties – temperature, density, pressure, and Mach number—in an extremely violent and irreversible fashion called a shock wave. The presence of shock waves, along with the compressibility effects of high-flow velocity fluids, is the central difference between the supersonic and subsonic aerodynamics regimes.

Hypersonic Flow

In aerodynamics, hypersonic speeds are speeds that are highly supersonic. In the 1970s, the term generally came to refer to speeds of Mach 5 (5 times the speed of sound) and above. The hypersonic regime is a subset of the supersonic regime. Hypersonic flow is characterized by high temperature flow behind a shock wave, viscous interaction, and chemical dissociation of gas.

Associated Terminology

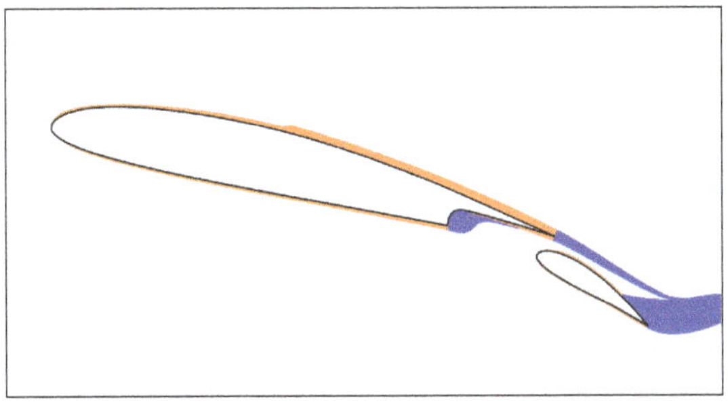

Different types flow analysis around an airfoil.

▯ Potential flow theory ▮ Boundary layer flow theory ▮ Turbulent wake analysis

The incompressible and compressible flow regimes produce many associated phenomena, such as boundary layers and turbulence.

Boundary Layers

The concept of a boundary layer is important in many problems in aerodynamics. The viscosity and fluid friction in the air is approximated as being significant only in this thin layer. This assumption makes the description of such aerodynamics much more tractable mathematically.

Turbulence

In aerodynamics, turbulence is characterized by chaotic property changes in the flow. These include low momentum diffusion, high momentum convection, and rapid variation of pressure and flow velocity in space and time. Flow that is not turbulent is called laminar flow.

Aerodynamic Force

The force exerted on a body whenever there is a relative velocity between the body and the air. There are only two basic sources of aerodynamic force: the pressure distribution and the frictional shear stress distribution exerted by the airflow on the body surface. The pressure exerted by the air at a point on the surface acts perpendicular to the surface at that point; and the shear stress, which is due to the frictional action of the air rubbing against the surface, acts tangentially to the surface at that point. The distribution of pressure and shear stress represent a distributed load over the surface. The net aerodynamic force on the body is due to the net imbalance between these distributed loads as they are summed (integrated) over the entire surface.

For purposes of discussion, it is convenient to consider the aerodynamic force on an airfoil. The net resultant aerodynamic force *R* acting through the center of pressure on the airfoil represents mechanically the same effect as that due to the actual pressure and shear stress loads distributed over the body surface. The velocity of the airflow V_∞ is called the free-stream velocity or the free-stream relative wind. By definition, the component of *R* perpendicular to the relative wind is the lift, *L*, and the component of *R* parallel to the relative wind is the drag *D*. The orientation of the body with respect to the direction of the free stream is given by the angle of attack, α. The magnitude of the aerodynamic force *R* is governed by the density & rgr;$_\infty$ and velocity of the free stream, the size of the body, and the angle of attack.

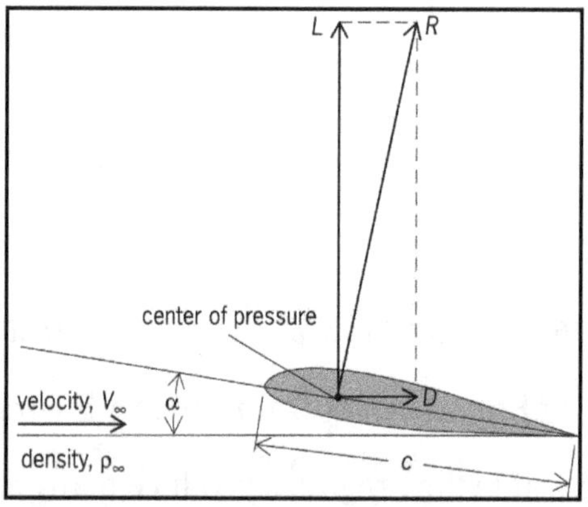

Resultant aerodynamic force (*R*), and its resolution into lift (*L*) and drag (*D*) components.

An important measure of aerodynamic efficiency is the ratio of lift to drag, *L*/*D*. The higher the value of *L*/*D*, the more efficient is the lifting action of the body. The value of *L*/*D* reaches a maximum, denoted by $(L/D)_{max}$, at a relatively low angle of attack. Beyond a certain angle the lift decreases with increasing α. In this region, the wing is said to be stalled. In the stall region the flow has separated from the top surface of the wing, creating a type of slowly recirculating dead-air region, which decreases the lift and substantially increases the drag.

Lift Force

Lift is the force that directly opposes the weight of an airplane and holds the airplane in the air. Lift is generated by every part of the airplane, but most of the lift on a normal airliner is generated by the wings. Lift is a mechanical aerodynamic force produced by the motion of the airplane through the air. Because lift is a force, it is a vector quantity, having both a magnitude and a direction associated with it. Lift acts through the center of pressure of the object and is directed perpendicular to the flow direction. There are several factors which affect the magnitude of lift.

How is Lift Generated?

There are many explanations for the generation of lift found in encyclopedias, in basic physics textbooks, and on Web sites. Unfortunately, many of the explanations are misleading and incorrect. Theories on the generation of lift have become a source of great controversy and a topic for heated arguments. To help you understand lift and its origins, a series of pages will describe the various theories and how some of the popular theories fail.

Lift occurs when a moving flow of gas is turned by a solid object. The flow is turned in one direction, and the lift is generated in the opposite direction, according to Newton's Third Law of action and reaction. Because air is a gas and the molecules are free to move about, any solid surface can deflect a flow. For an aircraft wing, both the upper and lower surfaces contribute to the flow turning. Neglecting the upper surface's part in turning the flow leads to an incorrect theory of lift.

No Fluid, No Lift

Lift is a mechanical force. It is generated by the interaction and contact of a solid body with a fluid (liquid or gas). It is not generated by a force field, in the sense of a gravitational field,or an electromagnetic field, where one object can affect another object without being in physical contact. For lift to be generated, the solid body must be in contact with the fluid: no fluid, no lift. The Space Shuttle does not stay in space because of lift from its wings but because of orbital mechanics related to its speed. Space is nearly a vacuum. Without air, there is no lift generated by the wings.

No Motion, No Lift

Lift is generated by the difference in velocity between the solid object and the fluid. There must be motion between the object and the fluid: no motion, no lift. It makes no difference whether the object moves through a static fluid, or the fluid moves past a static solid object. Lift acts perpendicular to the motion. Drag acts in the direction opposed to the motion.

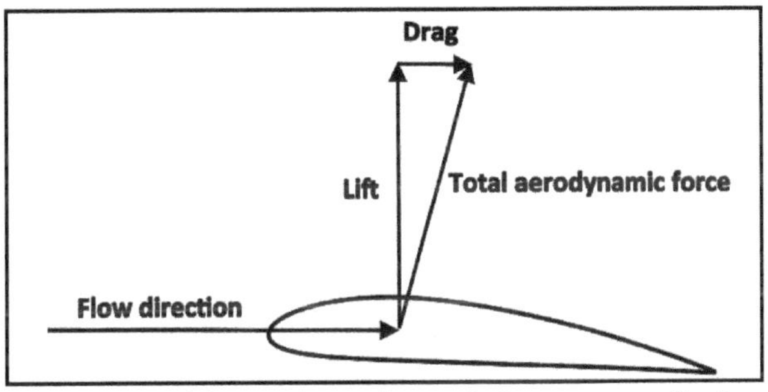

Lift is defined as the component of the total aerodynamic force perpendicular to the flow direction, and drag is the component parallel to the flow direction.

A fluid flowing past the surface of a body exerts a force on it. It makes no difference whether the fluid is flowing past a stationary body or the body is moving through a stationary volume of fluid. Lift is the component of this force that is perpendicular to the oncoming flow direction. Lift is always accompanied by a drag force, which is the component of the surface force parallel to the flow direction.

Lift is most commonly associated with the wings of fixed-wing aircraft, although it is more generally generated by many other streamlined bodies such as propellers, kites, helicopter rotors, racing car wings, maritime sails, and wind turbines in air, and by sailboat keels, ship's rudders, and hydrofoils in water. Lift is also exploited in the animal world, especially by birds, bats, and insects, and even in the plant world by the seeds of certain trees.

While the common meaning of the word "lift" assumes that lift opposes weight, lift can be in any direction with respect to gravity, since it is defined with respect to the direction of flow rather than to the direction of gravity. When an aircraft is cruising in straight and level flight, most of the lift opposes gravity. However, when an aircraft is climbing, descending, or banking in a turn the lift is tilted with respect to the vertical. Lift may also act as downforce in some aerobatic manoeuvres, or on the wing on a racing car. Lift may also be largely horizontal, for instance on a sailing ship.

The Lift discussed in this article is mainly in relation to airfoils, although marine hydrofoils and propellers share the same physical principles and work in the same way, despite differences between air and water such as density, compressibility, and viscosity.

Simplified Physical Explanations of Lift on an Airfoil

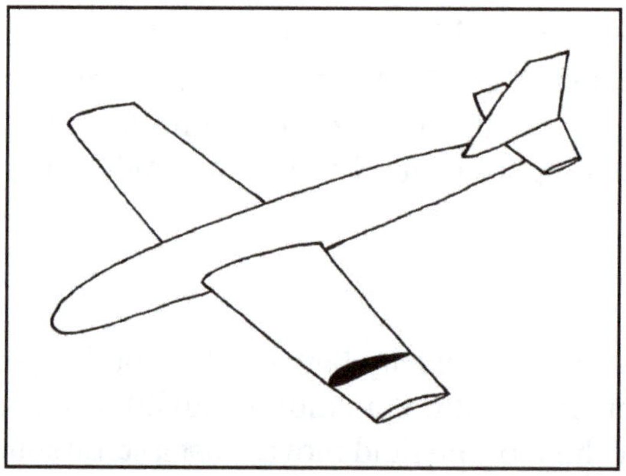

A cross-section of a wing defines an airfoil shape.

An airfoil is a streamlined shape that is capable of generating significantly more lift than drag. A flat plate can generate lift, but not as much as a streamlined airfoil, and with somewhat higher drag.

There are several ways to explain how an airfoil generates lift. Some are more complicated or more physically rigorous than others; some have been shown to be incorrect. For example, there are explanations based directly on Newton's laws of motion and explanations based on Bernoulli's principle. Either can be used to explain lift.

Flow Deflection and Newton's Laws

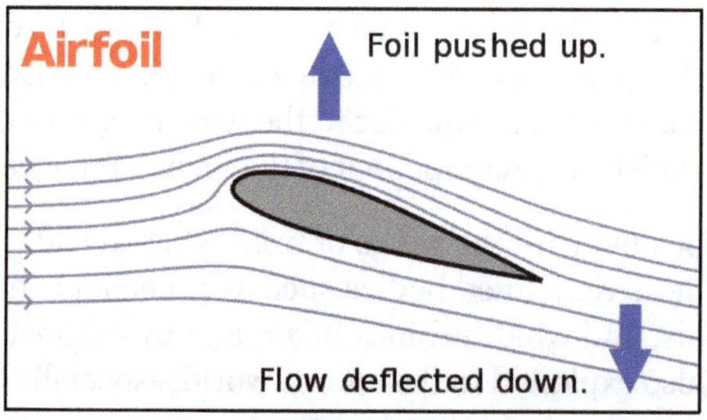

When an airfoil deflects air downwards, Newton's third law requires that the air must exert an equal upward reaction on the airfoil.

An airfoil generates lift by exerting a downward force on the air as it flows past. According to Newton's third law, the air must exert an equal and opposite (upward) force on the airfoil, which is Lift.

The airflow changes direction as it passes the airfoil and follows a path that is curved downward. According to Newton's second law, this change in flow direction requires a downward force applied to the air by the airfoil. Then Newton's third law requires the air to exert an upward force on the airfoil; thus a reaction force, Lift, is generated opposite to the directional change. In the case of an airplane wing, the wing exerts a downward force on the air and the air exerts an upward force on the wing.

The downward turning of the flow is not produced solely by the lower surface of the airfoil, and the air flow above the airfoil accounts for much of the downward-turning action.

Increased Flow Speed and Bernoulli's Principle

Bernoulli's principle states: "In a flow where no energy is being added or taken away, the sum of its various energies is a constant: consequently where the velocity increases the pressure decreases and vice versa." Thus, there is a direct mathematical relationship between the pressure and the speed, so if one knows the speed at all points within the airflow one can calculate the pressure, and vice versa. For any airfoil generating lift, there must be a pressure imbalance, i.e. lower average air pressure on the top than on the bottom. Bernoulli's principle states that this pressure difference must be accompanied by a speed difference.

Conservation of Mass

Streamlines and streamtubes around an airfoil generating lift. Note the narrower streamtubes above and the wider streamtubes below.

Starting with the flow pattern observed in both theory and experiments, the increased flow speed over the upper surface can be explained in terms of streamtube pinching and conservation of mass.

For incompressible flow, the rate of volume flow (e.g. volume units per minute) must be constant within each streamtube since matter is not created or destroyed. If a streamtube becomes narrower, the flow speed must increase in the narrower region to maintain the constant flow rate, to satisfy the principle of conservation of mass.

The upper streamtubes constrict as they flow up and around the airfoil. Conservation of mass says that the flow speed must increase as the stream tube area decreases. Similarly, the lower streamtubes expand and their flowrate slows.

From Bernoulli's principle, the pressure on the upper surface where the flow is moving faster is lower than the pressure on the lower surface where it is moving slower. This pressure difference creates a net aerodynamic force, pointing upward.

Limitations of the Simplified Explanations

Producing a lift force requires maintaining pressure differences in both the vertical and horizontal directions, and thus requires both downward turning of the flow and changes in flow speed consistent with Bernoulli's principle. The simplified explanations given above are therefore incomplete because they define lift in terms of only one or the other. And depending on the details, they have other shortcomings as well.

The explanation based on Flow deflection and Newton's laws is correct but is incomplete. It does not explain how the airfoil can impart downward turning to a much deeper swath of the flow than it actually touches. Further, it doesn't explain how the pressure differences in the horizontal direction are sustained. That is, it leaves out the Bernoulli part of the interaction.

Explanations based on Increased flow speed and Bernoulli's principle first try to establish that there is higher flow speed over the upper surface, but they fail to explain correctly what causes the flow to speed up:

- The Conservation of mass explanation that relies on narrowing of the streamtubes over the upper surface does not explain why the streamtubes change size. To see why the air flows the way it does requires more sophisticated analysis.
- Sometimes a geometrical argument is offered to demonstrate why the streamtubes change size: it is asserted that the top "obstructs" or "constricts" the air more than the bottom, hence narrower streamtubes. For conventional wings that are flat on the bottom and curved on top this makes some intuitive sense. But it does not explain how flat plates, symmetric airfoils, sailboat sails, or conventional airfoils flying upside down can generate lift, and attempts to calculate lift based on the amount of constriction do not predict experimental results.
- A common version that relies on equal-transit-time is simply wrong, as explained below under False explanation based on equal transit-time.

Bernoulli-only explanations imply that a speed difference arises from causes other than a pressure difference, and that the speed difference then leads to a pressure difference by Bernoulli's principle. This implied one-way causation is a misconception. The real cause-and-effect relationship between pressure and velocity is reciprocal. Finally, Bernoulli-only explanations don't explain how the pressure differences in the vertical direction are sustained. That is, they leave out the downward-turning part of the interaction.

Basic Attributes of Lift

Lift is a result of pressure differences and depends on angle of attack, airfoil shape, air density, and airspeed.

Pressure Differences

Pressure is the normal force per unit area exerted by the air on itself and on surfaces that it touches. The lift force is transmitted through the pressure, which acts perpendicular to the surface of the airfoil. The air maintains physical contact at all points. Thus, the net force manifests itself as pressure differences. The direction of the net force implies that the average pressure on the upper surface of the airfoil is lower than the average pressure on the underside.

These pressure differences arise in conjunction with the curved airflow. When a fluid follows a curved path, there is a pressure gradient perpendicular to the flow direction with higher pressure on the outside of the curve and lower pressure on the inside. This direct relationship between curved streamlines and pressure differences, sometimes called the streamline curvature theorem, was derived from Newton's second law by Leonhard Euler in 1754:

$$\frac{\mathrm{d}p}{\mathrm{d}R} = \rho\frac{v^2}{R}$$

The left side of this equation represents the pressure difference perpendicular to the fluid flow. On the right hand side ρ is the density, v is the velocity, and R is the radius of curvature. This formula shows that higher velocities and tighter curvatures create larger pressure differentials and that for straight flow ($R \rightarrow \infty$) the pressure difference is zero.

Angle of Attack

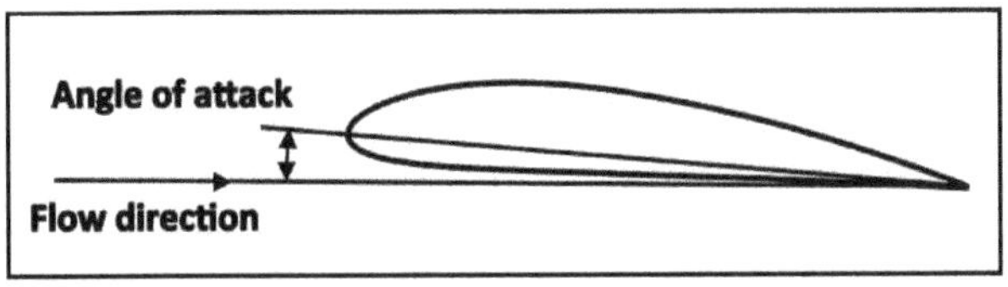

Angle of attack of an airfoil.

The angle of attack is the angle between the chord line of an airfoil and the oncoming airflow. A symmetrical airfoil will generate zero lift at zero angle of attack. But as the angle of attack increases, the air is deflected through a larger angle and the vertical component of the airstream velocity increases, resulting in more lift. For small angles a symmetrical airfoil will generate a lift force roughly proportional to the angle of attack.

As the angle of attack increases, the lift reaches a maximum at some angle; increasing the angle of attack beyond this critical angle of attack causes the upper-surface flow to separate from the wing; there is less deflection downward so the airfoil generates less lift. The airfoil is said to be stalled.

Airfoil Shape

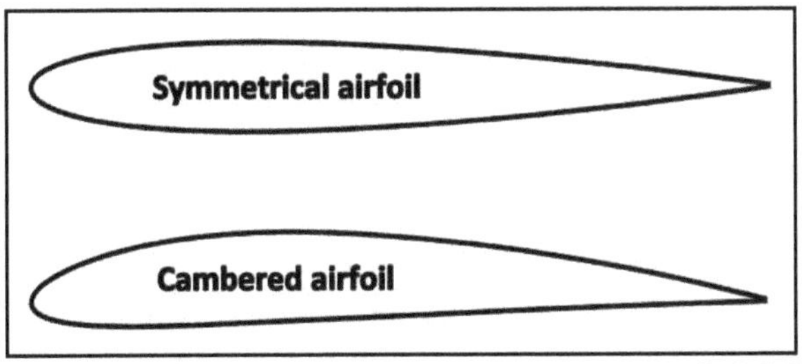

An airfoil with camber compared to a symmetrical airfoil.

The lift force depends on the shape of the airfoil, especially the amount of camber (curvature such that the upper surface is more convex than the lower surface, as illustrated at right). Increasing the camber generally increases lift.

Cambered airfoils will generate lift at zero angle of attack. When the chord line is horizontal, the trailing edge has a downward direction and since the air follows the trailing edge it is deflected downward. When a cambered airfoil is upside down, the angle of attack can be adjusted so that the lift force is upwards. This explains how a plane can fly upside down.

Flow Conditions

The ambient flow conditions which affect lift include the fluid density, viscosity and speed of flow. Density is affected by temperature, and by the medium's acoustic velocity - i.e. by compressibility effects.

Air Speed and Density

Lift is proportional to the density of the air and approximately proportional to the square of the flow speed. Lift also depends on the size of the wing, being generally proportional to the wing's area projected in the lift direction. In calculations it is convenient to quantify lift in terms of a "Lift coefficient" based on these factors.

Boundary Layer and Profile Drag

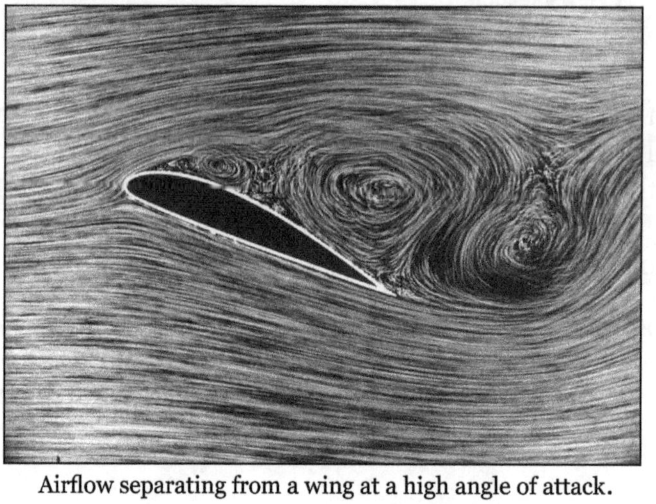

Airflow separating from a wing at a high angle of attack.

No matter how smooth the surface of an airfoil seems, any surface is rough on the scale of air molecules. Air molecules flying into the surface bounce off the rough surface in random directions superposed on their original velocity. The result is that when the air is viewed as if it were a continuous material, it is seen to be unable to slide along the surface, and the air's velocity relative to the airfoil decreased to nearly zero at the surface. goes to practically zero, something known as the no-slip condition. Because the air at the surface has near-zero velocity, and air away from the surface is moving, there is a thin boundary layer in which air close to the surface is subjected to a shearing motion. The air's viscosity resists the shearing, giving rise to a shear stress at the airfoil's surface called skin-friction drag. Over most of the surface of most airfoils, the boundary layer is naturally turbulent, which increases skin-friction drag.

Under usual flight conditions, the boundary layer remains attached to both the upper and lower surfaces all the way to the trailing edge, and its effect on the rest of the flow is modest. Compared to the predictions of inviscid-flow theory, in which there is no boundary layer, the attached boundary layer reduces the lift by a modest amount and modifies the pressure distribution somewhat, which results in a viscosity-related pressure drag over and above the skin-friction drag. The total of the skin-friction drag and the viscosity-related pressure drag is usually called the profile drag.

Stalling

An airfoil's maximum lift at a given airspeed is limited by boundary-layer separation. As the angle of attack is increased, a point is reached where the boundary layer can no longer remain attached to the upper surface. When the boundary layer separates, it leaves a region of recirculating flow above the upper surface, as illustrated in the flow-visualization photo at right. This is known as the *stall*, or *stalling*. At angles of attack above the stall, lift is significantly reduced, though it does not drop to zero. The maximum lift that can be achieved before stall, in terms of the lift coefficient, is generally less than 1.5 for single-element airfoils and can be more than 3.0 for airfoils with high-lift slotted flaps and leading-edge devices deployed.

Bluff Bodies

The flow around bluff bodies – i.e. without a streamlined shape, or stalling airfoils – may also generate lift, in addition to a strong drag force. This lift may be steady, or it may oscillate due to vortex shedding. Interaction of the object's flexibility with the vortex shedding may enhance the effects of fluctuating lift and cause vortex-induced vibrations. For instance, the flow around a circular cylinder generates a Kármán vortex street: vortices being shed in an alternating fashion from the cylinder's sides. The oscillatory nature of the flow produces a fluctuating lift force on the cylinder, even though the net (mean) force is negligible. The lift force frequency is characterised by the dimensionless Strouhal number, which depends on the Reynolds number of the flow.

For a flexible structure, this oscillatory lift force may induce vortex-induced vibrations. Under certain conditions – for instance resonance or strong spanwise correlation of the lift force – the resulting motion of the structure due to the lift fluctuations may be strongly enhanced. Such vibrations may pose problems and threaten collapse in tall man-made structures like industrial chimneys.

In the Magnus effect, a lift force is generated by a spinning cylinder in a freestream. Here the mechanical rotation acts on the boundary layer, causing it to separate at different locations on the two sides of the cylinder. The asymmetric separation changes the effective shape of the cylinder as far as the flow is concerned such that the cylinder acts like a lifting airfoil with circulation in the outer flow.

A More Comprehensive Physical Explanation

There are two main popular explanations: one based on downward deflection of the flow (Newton's laws), and one based on pressure differences accompanied by changes in flow speed (Bernoulli's principle). Either of these, by itself, correctly identifies some aspects of the lifting flow but leaves other important aspects of the phenomenon unexplained. A more comprehensive explanation involves both downward deflection and pressure differences (including changes in flow speed associated with the pressure differences).

Lift at the Airfoil Surface

The airfoil shape and angle of attack work together so that the airfoil exerts a downward force on the air as it flows past. According to Newton's third law, the air must then exert an equal and opposite (upward) force on the airfoil, which is the lift.

The net force exerted by the air occurs as a pressure difference over the airfoil's surfaces. Pressure in a fluid is always positive in an absolute sense, so that pressure must always be thought of as pushing, and never as pulling. The pressure thus pushes inward on the airfoil everywhere on both the upper and lower surfaces. The flowing air reacts to the presence of the wing by reducing the pressure on the wing's upper surface and increasing the pressure on the lower surface. The pressure on the lower surface pushes up harder than the reduced pressure on the upper surface pushes down, and the net result is upward lift.

The pressure difference which results in Lift acts directly on the airfoil surfaces; however, understanding how the pressure difference is produced requires understanding what the flow does over a wider area.

The Wider Flow around the Airfoil

An airfoil affects the speed and direction of the flow over a wide area, producing a pattern called a *velocity field*. When an airfoil produces lift, the flow ahead of the airfoil is deflected upward, the flow above and below the airfoil is deflected downward, and the flow behind the airfoil is deflected upward again, leaving the air far behind the airfoil in the same state as the oncoming flow far ahead. The flow above the upper surface is sped up, while the flow below the airfoil is slowed down. Together with the upward deflection of air in front and the downward deflection of the air immediately behind, this establishes a net circulatory component of the flow. The downward deflection and the changes in flow speed are pronounced and extend over a wide area, as can be seen in the image below. These differences in the direction and speed of the flow are greatest close to the airfoil and decrease gradually far above and below. All of these features of the velocity field also appear in theoretical models for lifting flows.

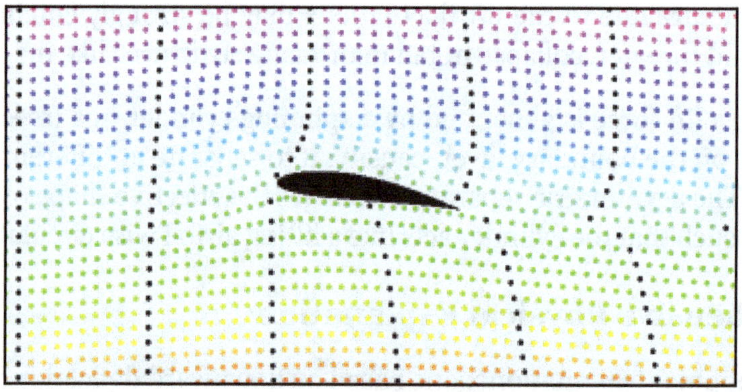

Flow around an airfoil: the dots move with the flow.

The black dots are on time slices, which split into two – an upper and lower part – at the leading edge. A marked speed difference between the upper-and lower-surface streamlines is shown most clearly in the image, with the upper markers arriving at the trailing edge long before the lower ones. Colors of the dots indicate streamlines.

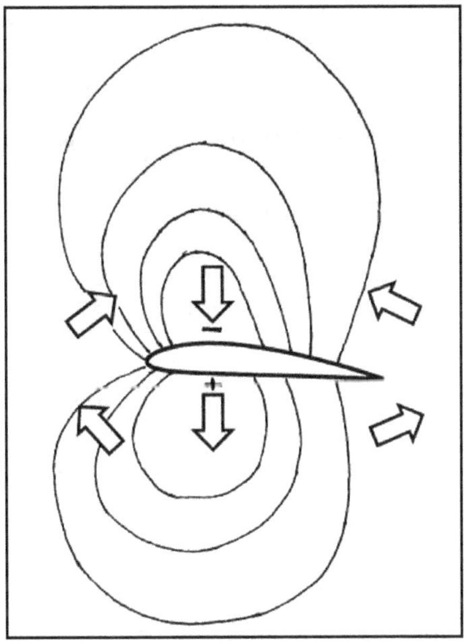

Pressure distribution with isobars around a lifting airfoil. The plus sign indicates pressure higher than ambient, and the minus sign indicates pressure lower than ambient (NOT negative pressure in the absolute sense). The block arrows indicate the directions of net forces on fluid parcels in different parts of the flowfield.

The pressure is also affected over a wide area, in a pattern of non-uniform pressure called a *pressure field*. When an airfoil produces lift, there is a diffuse region of low pressure above the airfoil, and usually a diffuse region of high pressure below, as illustrated by the isobars (curves of constant pressure) in the drawing. The pressure difference that acts on the surface is just part of this pressure field.

Mutual Interaction of Pressure Differences and Changes in Flow Velocity

The non-uniform pressure exerts forces on the air in the direction from higher pressure to lower pressure. The direction of the force is different at different locations around the airfoil, as indicated by the block arrows in the *pressure distribution with isobars* figure. Air above the airfoil is pushed toward the center of the low-pressure region, and air below the airfoil is pushed outward from the center of the high-pressure region.

According to *Newton's second law*, a force causes air to accelerate in the direction of the force. Thus the vertical arrows in the *pressure distribution with isobars* figure indicate that air above and below the airfoil is accelerated, or turned downward, and that the non-uniform pressure is thus the cause of the downward deflection of the flow visible in the flow animation. To produce this downward turning, the airfoil must have a positive angle of attack or have its rear portion curved downward as on an airfoil with camber. Note that the downward turning of the flow over the upper surface is the result of the air being pushed downward by higher pressure above it than below it. Some explanations that refer to the "Coandă effect" suggest that viscosity plays a key role in the downward turning, but this is false.

The arrows ahead of the airfoil indicate that the flow ahead of the airfoil is deflected upward, and the arrows behind the airfoil indicate that the flow behind is deflected upward again, after being deflected downward over the airfoil.

The arrows ahead of the airfoil and behind also indicate that air passing through the low-pressure region above the airfoil is sped up as it enters, and slowed back down as it leaves. Air passing through the high-pressure region below the airfoil sees the opposite - it is slowed down and then sped up. Thus the non-uniform pressure is also the cause of the changes in flow speed visible in the flow animation. The changes in flow speed are consistent with *Bernoulli's principle*, which states that in a steady flow without viscosity, lower pressure means higher speed, and higher pressure means lower speed.

Thus changes in flow direction and speed are directly caused by the non-uniform pressure. But this cause-and-effect relationship is not just one-way; it works in both directions simultaneously. The air's motion is affected by the pressure differences, but the existence of the pressure differences depends on the air's motion. The relationship is thus a mutual, or reciprocal, interaction: Air flow changes speed or direction in response to pressure differences, and the pressure differences are sustained by the air's resistance to changing speed or direction. A pressure difference can exist only if something is there for it to push against. In aerodynamic flow, the pressure difference pushes against the air's inertia, as the air is accelerated by the pressure difference. This is why the air's mass is part of the calculaqtion, and why Lift depends on air density.

Sustaining the pressure difference that exerts the lift force on the airfoil surfaces requires sustaining a pattern of non-uniform pressure in a wide area around the airfoil. This requires maintaining pressure differences in both the vertical and horizontal directions, and thus requires both downward turning of the flow and changes in flow speed according to Bernoulli's principle. The pressure differences and the changes in flow direction and speed sustain each other in a mutual interaction. The pressure differences follow naturally from Newton's second law and from the fact that flow along the surface follows the predominantly downward-sloping contours of the airfoil. And the fact that the air has mass is crucial to the interaction.

Simpler Explanations

Producing a lift force requires both downward turning of the flow and changes in flow speed consistent with Bernoulli's principle. Each of the simplified explanations given above in Simplified physical explanations of lift on an airfoil falls short by trying to explain lift in terms of only one or the other, thus explaining only part of the phenomenon and leaving other parts unexplained.

Quantifying Lift

Pressure Integration

When the pressure distribution on the airfoil surface is known, determining the total lift requires adding up the contributions to the pressure force from local elements of the surface, each with its own local value of pressure. The total lift is thus the integral of the pressure, in the direction perpendicular to the farfield flow, over the airfoil surface.

$$L = \oint p\mathrm{n} \cdot \mathrm{k}\, dS,$$

where:

- S is the projected (planform) area of the airfoil, measured normal to the mean airflow;

- n is the normal unit vector pointing into the wing;
- k is the vertical unit vector, normal to the freestream direction.

The above lift equation neglects the skin friction forces, which are small compared to the pressure forces.

By using the streamwise vector i parallel to the freestream in place of k in the integral, we obtain an expression for the pressure drag D_p (which includes the pressure portion of the profile drag and, if the wing is three-dimensional, the induced drag). If we use the spanwise vector j, we obtain the side force Y.

$$D_p = \oint p\mathrm{n} \cdot \mathrm{i}\, dS,$$

$$Y = \oint p\mathrm{n} \cdot \mathrm{j}\, dS.$$

The validity of this integration generally requires the airfoil shape to be a closed curve that is piecewise smooth.

Lift Coefficient

Lift depends on the size of the wing, being approximately proportional to the wing area. It is often convenient to quantify the lift of a given airfoil by its *lift coefficient* C_L, which defines its overall lift in terms of a unit area of the wing.

If the value of C_L for a wing at a specified angle of attack is given, then the lift produced for specific flow conditions can be determined:

$$L = \tfrac{1}{2}\rho v^2 S C_L$$

where

- L is the lift force;
- ρ is the air density;
- v is the velocity or true airspeed;
- S is the planform (projected) wing area;
- C_L is the lift coefficient at the desired angle of attack, Mach number, and Reynolds number.

Three-dimensional Flow

The flow around a three-dimensional wing involves significant additional issues, especially relating to the wing tips. For a wing of low aspect ratio, such as a typical delta wing, two-dimensional thories may provide a poor model and three-dimensional flow effects can dominate. Even for wings of high aspect ratio, the three-dimensional effects associated with finite span can affect the whole span, not just close to the tips.

Cross-section of an airplane wing-body combination showing the isobars of the three-dimensional lifting flow.

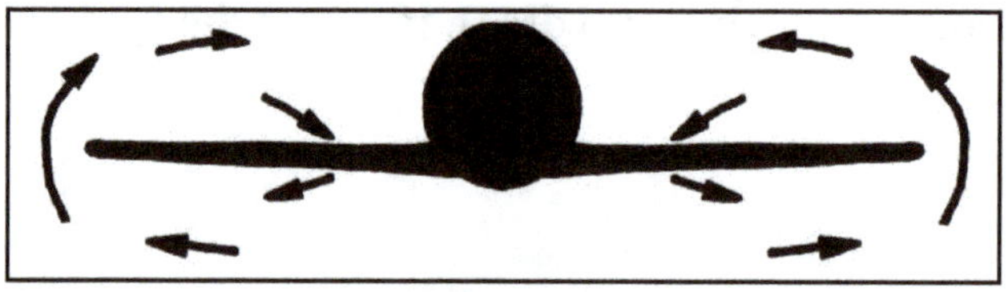

Cross-section of an airplane wing-body combination showing velocity vectors of the three-dimensional lifting flow

Wing Tips and Spanwise Distribution

The vertical pressure gradient at the wing tips causes air to flow sideways, out from under the wing then up and back over the upper surface. This reduces the pressure gradient at the wing tip, therefore also reducing lift. The lift tends to decrease in the spanwise direction from root to tip, and the pressure distributions around the airfoil sections change accordingly in the spanwise direction. Pressure distributions in planes perpendicular to the flight direction tend to look like the illustration at right. This spanwise-varying pressure distribution is sustained by a mutual interaction with the velocity field. Flow below the wing is accelerated outboard, flow outboard of the tips is accelerated upward, and flow above the wing is accelerated inboard, which results in a flow pattern.

There is more downward turning of the flow than there would be in a two-dimensional flow with the same airfoil shape and sectional lift, and a higher sectional angle of attack is required to achieve the same lift compared to a two-dimensional flow. The wing is effectively flying in a downdraft of its own making, as if the freestream flow were tilted downward, with the result that the total aerodynamic force vector is tilted backward slightly compared to what it would be in two dimensions. The additional backward component of the force vector is called lift-induced drag.

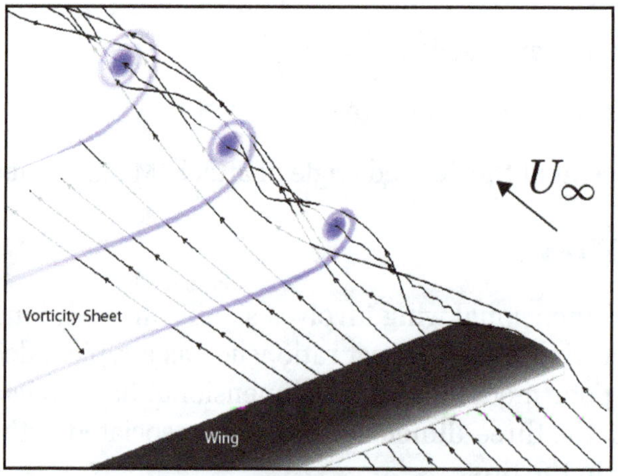

Euler computation of a tip vortex rolling up from the trailed vorticity sheet.

The difference in the spanwise component of velocity above and below the wing (between being in the inboard direction above and in the outboard direction below) persists at the trailing edge and into the wake downstream. After the flow leaves the trailing edge, this difference in velocity takes place across a relatively thin shear layer called a vortex sheet.

Horseshoe Vortex System

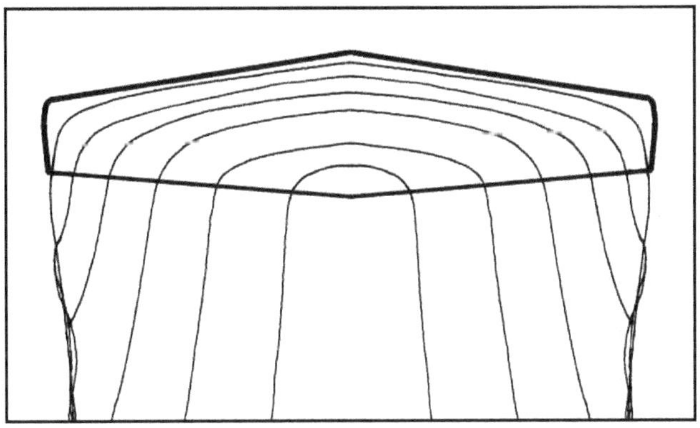

Planview of a wing showing the horseshoe vortex system.

The wingtip flow leaving the wing creates a tip vortex. As the main vortex sheet passes downstream from the trailing edge, it rolls up at its outer edges, merging with the tip vortices. The combination of the wingtip vortices and the vortex sheets feeding them is called the vortex wake.

In addition to the vorticity in the trailing vortex wake there is vorticity in the wing's boundary layer, called 'bound vorticity', which connects the trailing sheets from the two sides of the wing into a vortex system in the general form of a horseshoe. The horseshoe form of the vortex system was recognized by the British aeronautical pioneer Lanchester in 1907.

Given the distribution of bound vorticity and the vorticity in the wake, the Biot-Savart law (a vector-calculus relation) can be used to calculate the velocity perturbation anywhere in the field, caused by the lift on the wing. Approximate theories for the lift distribution and lift-induced drag of three-dimensional wings are based on such analysis applied to the wing's horseshoe vortex system. In these theories, the bound vorticity is usually idealized and assumed to reside at the camber surface inside the wing.

Because the velocity is deduced from the vorticity in such theories, some authors describe the situation to imply that the vorticity is the cause of the velocity perturbations, using terms such as "the velocity induced by the vortex," for example. But attributing mechanical cause-and-effect between the vorticity and the velocity in this way is not consistent with the physics. The velocity perturbations in the flow around a wing are in fact produced by the pressure field.

Manifestations of Lift in the Farfield

Integrated Force/Momentum Balance in Lifting Flows

The flow around a lifting airfoil must satisfy Newton's second law regarding conservation of momentum, both locally at every point in the flow field, and in an integrated sense over any extended region of the flow. For an extended region, Newton's second law takes the form of the *momentum theorem for a control volume*, where a control volume can be any region of the flow chosen for analysis. The momentum theorem states that the integrated force exerted at the boundaries of the control volume (a surface integral), is equal to the integrated time rate of change (material derivative) of the momentum of fluid parcels passing through the interior of the control volume. For a steady flow, this can be expressed in the form of the net surface integral of the flux of momentum through the boundary.

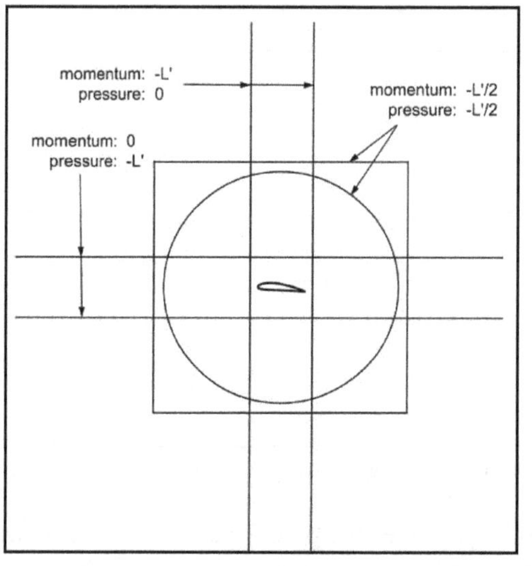

Control volumes of different shapes that have been used in analyzing the momentum balance in the 2D flow around a lifting airfoil. The airfoil is assumed to exert a downward force -L' per unit span on the air, and the proportions in which that force is manifested as momentum fluxes and pressure differences at the outer boundary are indicated for each different shape of control volume.

The lifting flow around a 2D airfoil is usually analyzed in a control volume that completely surrounds the airfoil, so that the inner boundary of the control volume is the airfoil surface, where the downward force per unit span $-L'$ is exerted on the fluid by the airfoil. The outer boundary is usually either a large circle or a large rectangle. At this outer boundary distant from the airfoil, the velocity and pressure are well represented by the velocity and pressure associated with a uniform flow plus a vortex, and viscous stress is negligible, so that the only force that must be integrated over the outer boundary is the pressure. The free-stream velocity is usually assumed to be horizontal, with lift vertically upward, so that the vertical momentum is the component of interest.

For the free-air case (no ground plane), the force $-L'$ exerted by the airfoil on the fluid is manifested partly as momentum fluxes and partly as pressure differences at the outer boundary, in proportions that depend on the shape of the outer boundary, as shown in the diagram at right. For a flat horizontal rectangle that is much longer than it is tall, the fluxes of vertical momentum through the front and back are negligible, and the lift is accounted for entirely by the integrated pressure differences on the top and bottom. For a square or circle, the momentum fluxes and pressure differences account for half the lift each. For a vertical rectangle that is much taller than it is wide, the unbalanced pressure forces on the top and bottom are negligible, and lift is accounted for entirely by momentum fluxes, with a flux of upward momentum that enters the control volume through the front accounting for half the lift, and a flux of downward momentum that exits the control volume through the back accounting for the other half.

Lift Reacted by Overpressure on the Ground under an Airplane

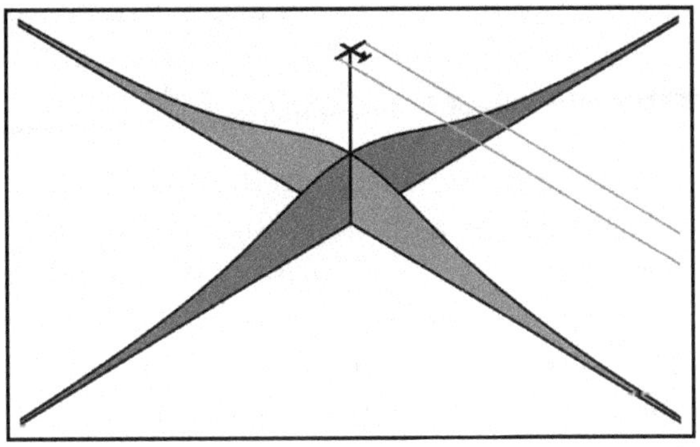

Illustration of the distribution of higher-than-ambient pressure on the ground under an airplane in flight.

An airfoil produces a pressure field in the surrounding air, as explained under "The wider flow around the airfoil" above. The pressure differences associated with this field die off gradually, becoming very small at large distances, but never disappearing altogether. Below the airplane, the pressure field persists as a positive pressure disturbance that reaches the ground, forming a pattern of slightly-higher-than-ambient pressure on the ground, as shown on the right. Although the pressure differences are very small far below the airplane, they are spread over a wide area and add up to a substantial force. For steady, level flight, the integrated force due to the pressure differences is equal to the total aerodynamic lift of the airplane and to the airplane's weight. According to Newton's third law, this pressure force exerted on the ground by the air is matched by an equal-and-opposite upward force exerted on the air by the ground, which offsets all of the downward force exerted on the air by the airplane. The net force due to the lift, acting on the atmosphere as a whole, is therefore zero, and thus there is no integrated accumulation of vertical momentum in the atmosphere, as was noted by Lanchester early in the development of modern aerodynamics.

Aerodynamic Drag

Drag is the aerodynamic force that opposes an aircraft's motion through the air. Drag is generated by every part of the airplane (even the engines). How is drag generated?

Drag is a mechanical force. It is generated by the interaction and contact of a solid body with a fluid (liquid or gas). It is not generated by a force field, in the sense of a gravitational field or an electromagnetic field, where one object can affect another object without being in physical contact. For drag to be generated, the solid body must be in contact with the fluid. If there is no fluid, there is no drag. Drag is generated by the difference in velocity between the solid object and the fluid. There must be motion between the object and the fluid. If there is no motion, there is no drag. It makes no difference whether the object moves through a static fluid or whether the fluid moves past a static solid object.

Drag is a force and is therefore a vector quantity having both a magnitude and a direction. Drag acts in a direction that is opposite to the motion of the aircraft. Lift acts perpendicular to the motion. There are many factors that affect the magnitude of the drag. Many of the factors also affect lift but there are some factors that are unique to aircraft drag.

The pressure distribution acting on a body's surface exerts normal forces on the body. Those forces can be summed and the component of that force that acts downstream represents the drag force, D_{pr}, due to pressure distribution acting on the body. The nature of these normal forces combines shock wave effects, vortex system generation effects, and wake viscous mechanisms.

The viscosity of the fluid has a major effect on drag. In the absence of viscosity, the pressure forces acting to retard the vehicle are canceled by a pressure force further aft that acts to push the vehicle forward; this is called pressure recovery and the result is that the drag is zero. That is to say, the work the body does on the airflow, is reversible and is recovered as there are no frictional effects to convert the flow energy into heat. Pressure recovery acts even in the case of viscous flow. Viscosity, however results in pressure drag and it is the dominant component of drag in the case of vehicles with regions of separated flow, in which the pressure recovery is fairly ineffective.

The friction drag force, which is a tangential force on the aircraft surface, depends substantially on boundary layer configuration and viscosity. The net friction drag, D_f, is calculated as the downstream projection of the viscous forces evaluated over the body's surface.

The sum of friction drag and pressure (form) drag is called viscous drag. This drag component is due to viscosity. In a thermodynamic perspective, viscous effects represent irreversible phenomena and, therefore, they create entropy. The calculated viscous drag D_v use entropy changes to accurately predict the drag force.

When the airplane produces lift, another drag component results. Induced drag, symbolized D_i, is due to a modification of the pressure distribution due to the trailing vortex system that accompanies the lift production. An alternative perspective on lift and drag is gained from considering the change of momentum of the airflow. The wing intercepts the airflow and forces the flow to move downward. This results in an equal and opposite force acting upward on the wing which is the lift force. The change of momentum of the airflow downward results in a reduction of the rearward momentum of the flow which is the result of a force acting forward on the airflow and applied by the wing to the air flow; an equal but opposite force acts on the wing rearward which is the induced drag. Induced drag tends to be the most important component for airplanes during take-off or landing flight. Another drag component, namely wave drag, D_w, results from shock waves in transonic and supersonic flight speeds. The shock waves induce changes in the boundary layer and pressure distribution over the body surface.

Lift-induced Drag

Lift-induced drag (also called induced drag) is drag which occurs as the result of the creation of lift on a three-dimensional lifting body, such as the wing or fuselage of an airplane. Induced drag consists primarily of two components: drag due to the creation of trailing vortices (vortex drag); and the presence of additional viscous drag (lift-induced viscous drag) that is not present when lift is zero. The trailing vortices in the flow-field, present in the wake of a lifting body, derive from the turbulent mixing of air from above and below the body which flows in slightly different directions as a consequence of creation of lift.

With other parameters remaining the same, as the lift generated by a body increases, so does the lift-induced drag. This means that as the wing's angle of attack increases (up to a maximum called the stalling angle), the lift coefficient also increases, and so too does the lift-induced drag. At the onset of stall, lift is abruptly decreased, as is lift-induced drag, but viscous pressure drag, a component of parasite drag, increases due to the formation of turbulent unattached flow in the wake behind the body.

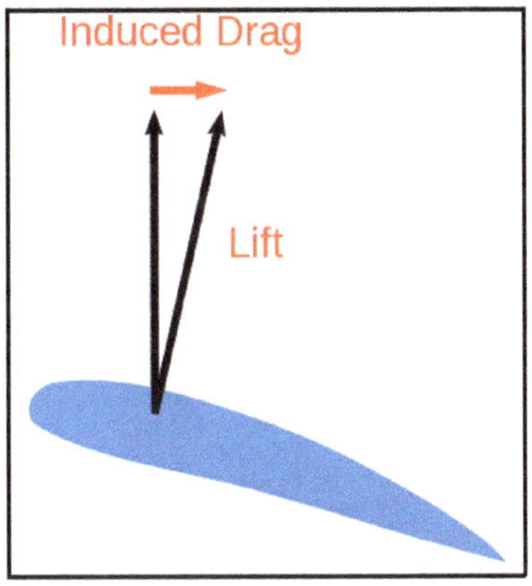

Induced drag *vs.* lift.

Parasitic Drag

Parasitic drag is drag caused by moving a solid object through a fluid. Parasitic drag is made up of multiple components including viscous pressure drag (form drag), and drag due to surface roughness (skin friction drag). Additionally, the presence of multiple bodies in relative proximity may incur so called interference drag, which is sometimes described as a component of parasitic drag.

In aviation, induced drag tends to be greater at lower speeds because a high angle of attack is required to maintain lift, creating more drag. However, as speed increases the angle of attack can be reduced and the induced drag decreases. Parasitic drag, however, increases because the fluid is flowing more quickly around protruding objects increasing friction or drag. At even higher speeds (transonic), wave drag enters the picture. Each of these forms of drag changes in proportion to the others based on speed. The combined overall drag curve therefore shows a minimum at some airspeed - an aircraft flying at this speed will be at or close to its optimal efficiency. Pilots will use this speed to maximize endurance (minimum fuel consumption), or maximize gliding range in the event of an engine failure.

Power Curve in Aviation

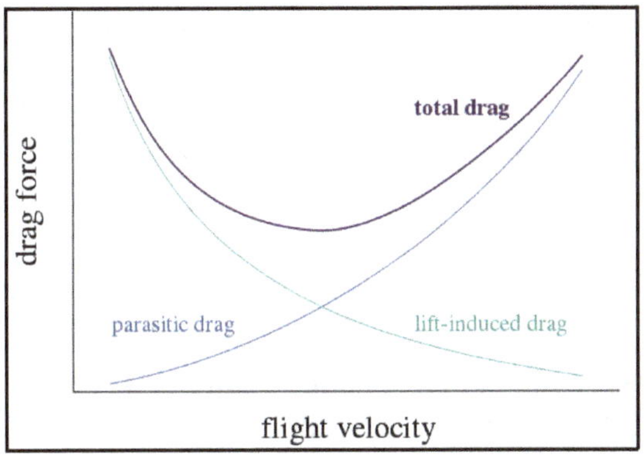

The *power curve*: form and induced drag *vs.* airspeed.

The interaction of parasitic and induced drag *vs.* airspeed can be plotted as a characteristic curve, illustrated here. In aviation, this is often referred to as the *power curve*, and is important to pilots because it shows that, below a certain airspeed, maintaining airspeed counterintuitively requires *more* thrust as speed decreases, rather than less. The consequences of being "behind the curve" in flight are important and are taught as part of pilot training. At the subsonic airspeeds where the "U" shape of this curve is significant, wave drag has not yet become a factor, and so it is not shown in the curve.

Wave Drag in Transonic and Supersonic Flow

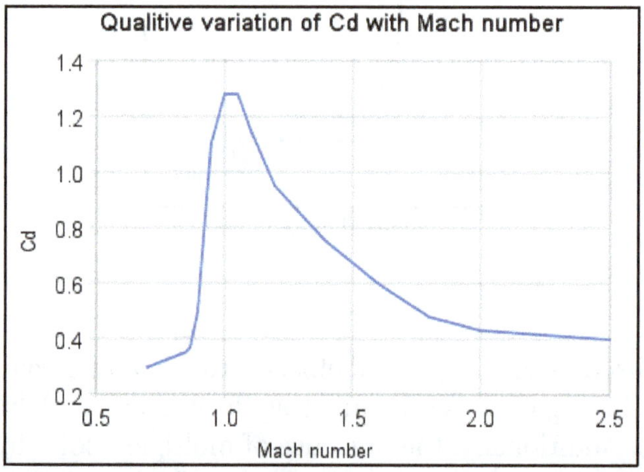

Qualitative variation in Cd factor with Mach number for aircraft.

Wave drag (also called compressibility drag) is drag that is created when a body moves in a compressible fluid and at speeds that are close to the speed of sound in that fluid. In aerodynamics, wave drag consists of multiple components depending on the speed regime of the flight.

In transonic flight (Mach numbers greater than about 0.8 and less than about 1.4), wave drag is the result of the formation of shockwaves in the fluid, formed when local areas of supersonic (Mach number greater than 1.0) flow are created. In practice, supersonic flow occurs on bodies traveling well below the speed of sound, as the local speed of air increases as it accelerates over the body to speeds above Mach 1.0. However, full supersonic flow over the vehicle will not develop until well past Mach 1.0. Aircraft flying at transonic speed often incur wave drag through the normal course of operation. In transonic flight, wave drag is commonly referred to as transonic compressibility drag. Transonic compressibility drag increases significantly as the speed of flight increases towards Mach 1.0, dominating other forms of drag at those speeds.

In supersonic flight (Mach numbers greater than 1.0), wave drag is the result of shockwaves present in the fluid and attached to the body, typically oblique shockwaves formed at the leading and trailing edges of the body. In highly supersonic flows, or in bodies with turning angles sufficiently large, unattached shockwaves, or bow waves will instead form. Additionally, local areas of transonic flow behind the initial shockwave may occur at lower supersonic speeds, and can lead to the development of additional, smaller shockwaves present on the surfaces of other lifting bodies, similar to those found in transonic flows. In supersonic flow regimes, wave drag is commonly separated into two components, supersonic lift-dependent wave drag and supersonic volume-dependent wave drag.

The closed form solution for the minimum wave drag of a body of revolution with a fixed length was found by Sears and Haack, and is known as the Sears-Haack Distribution. Similarly, for a fixed volume, the shape for minimum wave drag is the Von Karman Ogive.

Busemann's Biplane is not, in principle, subject to wave drag at all when operated at its design speed, but is incapable of generating lift.

Aircraft Drag Reduction

Every aircraft builder is interested in improving the speed of his airplane, or at least reducing its fuel consumption so that he/she gets the highest mileage out of every gallon or liter of fuel in the tanks or the lowest hourly rate.

Basically the form of the aircraft defines how much drag it generates and, in unaccelerated flight, how much thrust the engine propeller combination will need to produce to keep up the speed. The math is simple: this costs fuel.

We as the builder can try to keep the weight as low as possible (more lift results in more induced drag) but by paying attention while we construct the airframe we can also reduce parasite drag by streamlining parts as much as possible.

And by reducing the amount of drag (parasite) and keeping the empty weight to a minimum we consequently reduce fuel consumption and we can increase the payload and/or the range of our aircraft. We therefore have increased the efficiency of the aircraft/engine combination and the amount of money we can spent flying.

Some Aerodynamics

For an aircraft in unaccelerated level flight, the thrust is equal to the drag of the aircraft, remember Newton's first law. And we also know that Thrust (Drag) = Horsepower × propeller efficiency. Both variables thrust and drag are expressed in either pounds or kilograms.

This formula tells us that if you increase the thrust or decrease the drag, or even do both, the aircraft will go faster. But there is a mathematical relationship between power, drag and speed. The speed increases with the cube root of the power increase or in formula form: Speed increase = Power increase$^{1/3}$.

Take this example to clarify: your aircraft has a IAS of 100 kts on a 150 bhp Lycoming engine and you want to know how fast it will fly with a 180 bhp Lycoming. The increase in power is 1,2 (results from 180/150). The cube root of that increase ($1{,}2^{1/3}$) equals 1,063. Thus the increase in speed is 100 × 1,063 = 106,3 Kts. In reality you will not go that much faster by increasing engine power alone.

Higher Speeds

Drag also increases with the square of the speed. So if your aircraft total drag was 200 lbs at 100 Kts then increasing the speed to 106,3 Kts the total drag would end up to be $(106{,}3 / 100)^2 \times 200$ lbs = 226 lbs. That totals to 13 % more drag from a 6,3 % increase in speed.

We can conclude that drag reduction is the single best way to get additional speed from a certain design, and the faster the airplane, the more important each little improvement becomes.

Lift-induced Drag

Induced Drag is an inevitable consequence of lift and is produced by the passage of an aerofoil (e.g. wing or tailplane) through the air. Air flowing over the top of a wing tends to flow inwards because the decreased pressure over the top surface is less than the pressure outside the wing tip. Below the wing, the air flows outwards because the pressure below the wing is greater than that outside the wing tip. The direct consequence of this, as far as the wing tips are concerned, is that there is a continual spilling of air upwards around the wing tip a phenomenon called 'tip effect' or 'end effect'. One way to appreciate why a high aspect ratio for a wing is better than a low one is that with a high aspect ratio, the proportion of air which moves in this way is reduced and therefore more of it generates lift.

For the wing more generally, the streams of air from above and below the wing are flowing at an angle to each other as they meet along the trailing edge of the wing. They combine to form vortices which, when viewed from the rear, rotate clockwise from the left wing and counter clockwise from the right. The tendency is for these vortices to move outwards towards the wing tip joining together as they do so. By the time the wing tip is reached, one large wig tip vortex has formed and is shed.

Most of these vortices are of course completely invisible but, in very humid air, the central core of a vortex may become visible because the air pressure within its centre has reduced - and has therefore cooled - sufficiently for condensation to occur. A higher wing loading in a turn will also increase the strength - and the degree of reduced pressure - so that visible vortex cores are even more likely during turns. If close up to these vortices.

Most of the air flowing off the top of a wing - 'downwash' - continues more or less horizontally towards the empennage because it is balanced by a corresponding upwash in front of the wing leading edge. In contrast, the upwards air movement which leads to vortex 'consolidation' at the tip is just outside the tip whereas the corresponding downward movement is just at the extremity of the wingspan so that the net direction of airflow past the wing is downwards. The lift created by the wing - which is by definition at right angles to the airflow, is therefore inclined slightly backwards and thus 'contributes' drag - induced drag.

Source of Induced Drag

The total aerodynamic force acting on a body is usually thought of as having two components, lift and drag. By definition, the component of force parallel to the oncoming flow is called drag; and the component perpendicular to the oncoming flow is called lift. At practical angles of attack the lift greatly exceeds the drag.

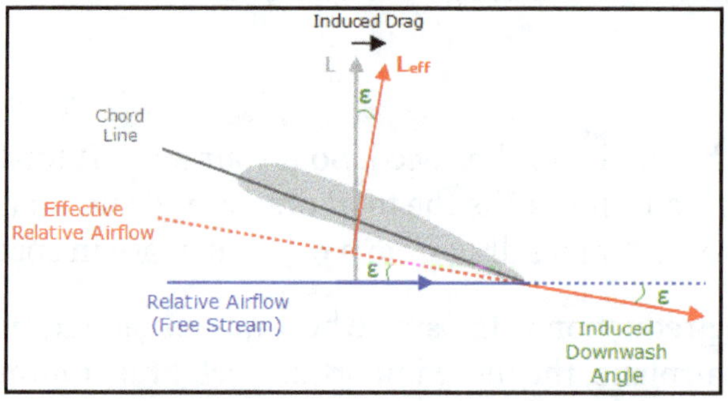

Induced drag is related to the amount of induced downwash in the vicinity of the wing.

The grey vertical line labeled "L" is perpendicular to the free stream and indicates the orientation of the lift on the wing. The red vector labeled "L_{eff}" is perpendicular to the actual airflow in the vicinity of the wing; it represents the lift on the airfoil section in two–dimensional flow at the same angle of attack. The lift generated by the wing has been tilted rearwards through an angle equal to the angle of the downwash in three-dimensional flow. The component of "L_{eff}" parallel to the free stream is the induced drag on the wing.

Lift is produced by the changing direction of the flow around a wing. The change of direction results in a change of velocity (even if there is no speed change, just as seen in uniform circular motion), which is an acceleration. To change the direction of the flow therefore requires that a force be applied to the fluid; lift is simply the reaction force of the fluid acting on the wing.

To produce lift, air below the wing is at a higher pressure than the air pressure above the wing. On a wing of finite span, this pressure difference causes air to flow from the lower surface wing root, around the wingtip, towards the upper surface wing root. This spanwise flow of air combines with chordwise flowing air, causing a change in speed and direction, which twists the airflow and produces vortices along the wing trailing edge. The vortices created are unstable, and they quickly combine to produce wingtip vortices. The resulting vortices change the speed and direction of the airflow behind the trailing edge, deflecting it downwards, and thus inducing downwash behind the wing.

Wingtip vortices modify the airflow around a wing, reducing wing's ability to generate lift, so that it requires a higher angle of attack for the same lift, which tilts the total aerodynamic force rearwards and increases the drag component of that force. The angular deflection is small and has little effect on the lift. However, there is an increase in the drag equal to the product of the lift force and the angle through which it is deflected. Since the deflection is itself a function of the lift, the additional drag is proportional to the square of the lift.

Reducing Induced Drag

According to the equations below, a wing of infinite aspect ratio (wingspan/chord length) and constant airfoil section would produce no induced drag. The characteristics of such a wing can be measured on a section of wing spanning the width of a wind tunnel, since the walls block spanwise flow and create what is effectively two-dimensional flow.

A rectangular planform wing produces stronger wingtip vortices than does a tapered or elliptical wing, therefore many modern wings are tapered. However, an elliptical planform is more efficient as the induced downwash (and therefore the effective angle of attack) is constant across the whole of the wingspan. Few aircraft have this planform because of manufacturing complications — the most famous examples being the World War II Spitfire and Thunderbolt. Tapered wings with straight leading and trailing edges can approximate to elliptical lift distribution. Typically, straight edged non-tapered wings produce 5%, and tapered wings produce 1-2% more induced drag than an elliptical wing.

Similarly, for a given wing area, a high aspect ratio wing will produce less induced drag than a wing of low aspect ratio because there is less air disturbance at the tip of a longer, thinner wing. Induced drag can therefore be said to be inversely proportional to aspect ratio. The lift distribution may also be modified by the use of washout, a spanwise twist of the wing to reduce the incidence towards the wingtips, and by changing the airfoil section near the wingtips. This allows more lift to be generated nearer the wing root and less towards the wingtip, which causes a reduction in the strength of the wingtip vortices.

Some early aircraft had fins mounted on the tips of the tailplane which served as endplates. More recent aircraft have wingtip mounted winglets to reduce the intensity of wingtip vortices. Wingtip mounted fuel tanks may also provide some benefit, by preventing the spanwise flow of air around the wingtip.

Calculation of Induced Drag

For a *planar* wing with an elliptical lift distribution, induced drag can be calculated as follows:

$$D_i = \frac{L^2}{\frac{1}{2}\rho V^2 \pi b^2},$$

where

L is the lift,

ρ is the air density,

V is the true airspeed, and

b is the wingspan.

From this equation it is clear that the induced drag decreases with flight speed and with wingspan. Deviation from the non-planar wing with elliptical lift distribution are taken into account by dividing the induced drag by the span efficiency factor e.

To compare with other sources of drag, it can be convenient to express this equation in terms of lift and drag coefficients:

$$C_{D,i} = \frac{D_i}{\frac{1}{2}\rho V^2 S} = \frac{C_L^2}{\pi e A\!\!\!R}, \text{ where}$$

$$C_L = \frac{L}{\frac{1}{2}\rho V^2 S}$$

and

$$A\!\!\!R = \frac{b^2}{S} \text{ is the aspect ratio,}$$

S is a reference wing area.

This indicates how high aspect ratio wings are beneficial to flight efficiency. With C_L being a function of angle of attack, induced drag increases as the angle of attack increases.

The above equation can be derived using Prandtl's lifting-line theory. Similar methods can also be used to compute the minimum induced drag for non-planar wings or for arbitrary lift distributions.

Combined Effect with other Drag Sources

Induced drag must be added to the parasitic drag to find the total drag. Since induced drag is inversely proportional to the square of the airspeed (at a given lift) whereas parasitic drag is proportional to the square of the airspeed, the combined overall drag curve shows a minimum at some airspeed - the minimum drag speed (V_{MD}). An aircraft flying at this speed is operating at its optimal aerodynamic efficiency. According to the above equations, the speed for minimum drag occurs at the speed where the induced drag is equal to the parasitic drag. This is the speed at which for unpowered aircraft, optimum glide angle is achieved. This is also the speed for greatest range (although V_{MD} will decrease as the plane consumes fuel and becomes lighter). The speed for greatest range (i.e., distance travelled) is the speed at which a straight line from the origin is tangent to the fuel flow rate curve. The curve of range versus airspeed is normally very flat and it is customary to operate at the speed for 99% best range since this gives about 5% greater speed for only 1% less range. (Of course, flying higher where the air is thinner will raise the speed at which minimum drag occurs, and so permits a faster voyage for the same amount of fuel. If the plane is flying at the maximum permissible speed, then there is an altitude at which the air density will be what is needed to keep it aloft while flying at the angle of attack that minimizes the drag. The optimum altitude at maximum speed, and the optimum speed at maximum altitude, may change during the flight as the plane becomes lighter).

The speed for maximum endurance (i.e., time in the air) is the speed for minimum fuel flow rate, and is less than the speed for greatest range. The fuel flow rate is calculated as the product of the power required and the engine specific fuel consumption (fuel flow rate per unit of power). The power required is equal to the drag times the speed.

Center of Pressure

As an object moves through a fluid, the velocity of the fluid varies around the surface of the object. The variation of velocity produces a variation of pressure on the surface of the object as shown by the the thin red lines on the figure. Integrating the pressure times the surface area around the body determines the aerodynamic force on the object. We can consider this single force to act through the average location of the pressure on the surface of the object. We call the average location of the pressure variation the center of pressure in the same way that we call the average location of the weight of an object the center of gravity. The aerodynamic force can then be resolved into two components, lift and drag, which act through the center of pressure in flight.

Determining the center of pressure is very important for any flying object. To trim an airplane, or to provide stability for a model rocket or a kite, it is necessary to know the location of the center of pressure of the entire aircraft. How do engineers determine the location of the center of pressure for an aircraft which they are designing?

In general, determining the center of pressure (cp) is a very complicated procedure because the pressure changes around the object. Determining the center of pressure requires the use of calculus and a knowledge of the pressure distribution around the body. We can characterize the pressure variation around the surface as a function p(x) which indicates that the pressure depends on the distance x from a reference line usually taken as the leading edge of the object. If we can determine the form of the function, there are methods to perform a calculus integration of the equation. We will use the symbols "S[ ]dx" to denote the integration of a continuous function. Then the center of pressure can be determined from:

cp = (S[x * p(x)]dx) / (S[p(x)]dx)

If we don't know the actual functional form, we can numerically integrate the equation using a spreadsheet by dividing the distance into a number of small distance segments and determining the average value of the pressure over that small segment. Taking the sum of the average value times the distance times the distance segment divided by the sum of the average value times the distance segment will produce the center of pressure.

Aircraft Aerodynamics

A stable configuration is desirable not only in sailing, but in aircraft design as well. Aircraft design therefore borrowed the term center of pressure. And like a sail, a rigid non-symmetrical airfoil not only produces lift, but a moment. The center of pressure of an aircraft is the point where all of the aerodynamic pressure field may be represented by a single force vector with no moment. A similar idea is the aerodynamic center which is the point on an airfoil where the pitching moment produced by the aerodynamic forces is constant with angle of attack.

The aerodynamic center plays an important role in analysis of the longitudinal static stability of all flying machines. It is desirable that when the pitch angle and angle of attack of an aircraft are disturbed (by, for example turbulence) that the aircraft returns to its original trimmed pitch angle and angle of attack without a pilot or autopilot changing the control surface deflection. For an aircraft to return towards its trimmed attitude, without input from a pilot or autopilot, it must have positive longitudinal static stability.

Missile Aerodynamics

Missiles typically do not have a preferred plane or direction of maneuver and thus have symmetric airfoils. Since the center of pressure for symmetric airfoils is relatively constant for small angle of attack, missile engineers typically speak of the complete center of pressure of the entire vehicle for stability and control analysis. In missile analysis, the center of pressure is typically defined as the center of the additional pressure field due to a change in the angle of attack off of the trim angle of attack.

For unguided rockets the trim position is typically zero angle of attack and the center of pressure is defined to be the center of pressure of the resultant flow field on the entire vehicle resulting from a very small angle of attack (that is, the center of pressure in the limit as angle of attack goes to zero). For positive stability in missiles, the total vehicle center of pressure defined as given above must be further from the nose of the vehicle than the center of gravity. In missiles at lower angles of attack, the contributions to the center of pressure are dominated by the nose, wings, and fins. The normalized normal force coefficient derivative with respect to the angle of attack of each component multiplied by the location of the center of pressure can be used to compute a centroid representing the total center of pressure. The center of pressure of the added flow field is behind the center of gravity and the additional force "points" in the direction of the added angle of attack; this produces a moment that pushes the vehicle back to the trim position.

In guided missiles where the fins can be moved to trim the vehicles in different angles of attack, the center of pressure is the center of pressure of the flow field at that angle of attack for the undeflected fin position. This is the center of pressure of any small change in the angle of attack (as defined above). Once again for positive static stability, this definition of center of pressure requires that the center of pressure be further from the nose than the center of gravity. This ensures that any increased forces resulting from increased angle of attack results in increased restoring moment to drive the missile back to the trimmed position. In missile analysis, positive static margin implies that the complete vehicle makes a restoring moment for any angle of attack from the trim position.

Movement of Center of Pressure for Aerodynamic Fields

The center of pressure on a symmetric airfoil typically lies close to 25% of the chord length behind the leading edge of the airfoil. (This is called the "quarter-chord point"). For a symmetric airfoil, as angle of attack and lift coefficient change, the center of pressure does not move. It remains around the quarter-chord point for angles of attack below the stalling angle of attack. The role of center of pressure in the control characterization of aircraft takes a different form than in missiles.

On a cambered airfoil the center of pressure does not occupy a fixed location. For a conventionally cambered airfoil, the center of pressure lies a little behind the quarter-chord point at maximum lift coefficient (large angle of attack), but as lift coefficient reduces (angle of attack reduces) the center of pressure moves toward the rear. When the lift coefficient is zero an airfoil is generating no lift but a conventionally cambered airfoil generates a nose-down pitching moment, so the location of the center of pressure is an infinite distance behind the airfoil.

For a reflex-cambered airfoil, the center of pressure lies a little ahead of the quarter-chord point at maximum lift coefficient (large angle of attack), but as lift coefficient reduces (angle of attack reduces) the center of pressure moves forward. When the lift coefficient is zero an airfoil is generating no lift but a reflex-cambered airfoil generates a nose-up pitching moment, so the location of the center of pressure is an infinite distance ahead of the airfoil. This direction of movement of the center of pressure on a reflex-cambered airfoil has a stabilising effect.

The way the center of pressure moves as lift coefficient changes makes it difficult to use the center of pressure in the mathematical analysis of longitudinal static stability of an aircraft. For this reason, it is much simpler to use the aerodynamic center when carrying out a mathematical analysis. The aerodynamic center occupies a fixed location on an airfoil, typically close to the quarter-chord point.

The aerodynamic center is the conceptual starting point for longitudinal stability. The horizontal stabilizer contributes extra stability and this allows the center of gravity to be a small distance aft of the aerodynamic center without the aircraft reaching neutral stability. The position of the center of gravity at which the aircraft has neutral stability is called the neutral point.

Thrust

Thrust is the force which moves an aircraft through the air. Thrust is used to overcome the drag of an airplane, and to overcome the weight of a rocket. Thrust is generated by the engines of the aircraft through some kind of propulsion system.

Thrust is a mechanical force, so the propulsion system must be in physical contact with a working fluid to produce thrust. Thrust is generated most often through the reaction of accelerating a mass of gas. Since thrust is a force, it is a vector quantity having both a magnitude and a direction. The engine does work on the gas and accelerates the gas to the rear of the engine; the thrust is generated in the opposite direction from the accelerated gas. The magnitude of the thrust depends on the amount of gas that is accelerated and on the difference in velocity of the gas through the engine.

To accelerate the gas, we have to expend energy. The energy is generated as heat by the combustion of some fuel. The thrust equation describes how the acceleration of the gas produces a force. The type of propulsion system used on an aircraft may vary from airplane to airplane and each device produces thrust in a slightly different way.

Examples

A fixed-wing aircraft generates forward thrust when air is pushed in the direction opposite to flight. This can be done in several ways including by the spinning blades of a propeller, or a rotating fan pushing air out from the back of a jet engine, or by ejecting hot gases from a rocket engine. The forward thrust is proportional to the mass of the airstream multiplied by the difference in velocity of the airstream. Reverse thrust can be generated to aid braking after landing by reversing the pitch of variable-pitch propeller blades, or using a thrust reverser on a jet engine. Rotary wing aircraft and thrust vectoring V/STOL aircraft use engine thrust to support the weight of the aircraft, and vector sum of this thrust fore and aft to control forward speed.

A rocket is propelled forward by a thrust force equal in magnitude, but opposite in direction, to the time-rate of momentum change of the exhaust gas accelerated from the combustion chamber through the rocket engine nozzle. This is the exhaust velocity with respect to the rocket, times the time-rate at which the mass is expelled, or in mathematical terms:

$$T = v\frac{dm}{dt}$$

Where T is the thrust generated (force), $\frac{dm}{dt}$ is the rate of change of mass with respect to time (mass flow rate of exhaust), and v is the speed of the exhaust gases measured relative to the rocket. For vertical launch of a rocket the initial thrust at liftoff must be more than the weight.

Each of the three Space Shuttle Main Engines could produce a thrust of 1.8 MN, and each of the Space Shuttle's two Solid Rocket Boosters 14.7 MN, together 29.4 MN.

By contrast, the simplified Aid For EVA Rescue (SAFER) has 24 thrusters of 3.56 N each.

In the air-breathing category, the AMT-USA AT-180 jet engine developed for radio-controlled aircraft produce 90 N (20 lbf) of thrust. The GE90-115B engine fitted on the Boeing 777-300ER, recognized by the Guinness Book of World Records as the "World's Most Powerful Commercial Jet Engine," has a thrust of 569 kN (127,900 lbf).

Concepts

Thrust to Power

The power needed to generate thrust and the force of the thrust can be related in a non-linear way. In general, $P^2 \propto T^3$. The proportionality constant varies, and can be solved for a uniform flow:

$$\frac{dm}{dt} = \rho A v$$

$$T = \frac{dm}{dt} v, P = \frac{1}{2} \frac{dm}{dt} v^2$$

$$T = \rho A v^2, P = \frac{1}{2} \rho A v^3$$

$$P^2 = \frac{T^3}{4 \rho A}$$

Note that these calculations are only valid for when the incoming air is accelerated from a standstill - for example when hovering.

The inverse of the proportionality constant, the "efficiency" of an otherwise-perfect thruster, is proportional to the area of the cross section of the propelled volume of fluid (A) and the density of the fluid (ρ). This helps to explain why moving through water is easier and why aircraft have much larger propellers than watercraft.

Thrust to Propulsive Power

A very common question is how to contrast the thrust rating of a jet engine with the power rating of a piston engine. Such comparison is difficult, as these quantities are not equivalent. A piston engine does not move the aircraft by itself (the propeller does that), so piston engines are usually rated by how much power they deliver to the propeller. Except for changes in temperature and air pressure, this quantity depends basically on the throttle setting.

A jet engine has no propeller, so the propulsive power of a jet engine is determined from its thrust as follows. Power is the force (F) it takes to move something over some distance (d) divided by the time (t) it takes to move that distance:

$$P = F \frac{d}{t}$$

In case of a rocket or a jet aircraft, the force is exactly the thrust (T) produced by the engine. If the rocket or aircraft is moving at about a constant speed, then distance divided by time is just speed, so power is thrust times speed:

$$P = Tv$$

This formula looks very surprising, but it is correct: the *propulsive power* (or *power available*) of a jet engine increases with its speed. If the speed is zero, then the propulsive power is zero. If a jet aircraft is at full throttle but attached to a static test stand, then the jet engine produces no propulsive power, however thrust is still produced. Compare that to a piston engine. The combination piston engine–propeller also has a propulsive power with exactly the same formula, and it will also be zero at zero speed -- but that is for the engine–propeller set. The engine alone will continue to produce its rated power at a constant rate, whether the aircraft is moving or not.

Now, imagine the strong chain is broken, and the jet and the piston aircraft start to move. At low speeds:

> The piston engine will have constant 100% power, and the propeller's thrust will vary with speed
> The jet engine will have constant 100% thrust, and the engine's power will vary with speed.

Excess Thrust

If a powered aircraft is generating thrust T and experiencing drag D, the difference between the two, T − D, is termed the excess thrust. The instantaneous performance of the aircraft is mostly dependent on the excess thrust.

Excess thrust is a vector and is determined as the vector difference between the thrust vector and the drag vector.

Centre of Thrust

The centre of thrust for an object is an average point at which the total thrust may be considered to apply. It may differ from the centre of gravity.

Boundary Layer

As an object moves through a fluid, or as a fluid moves past an object, the molecules of the fluid near the object are disturbed and move around the object. Aerodynamic forces are generated between the fluid and the object. The magnitude of these forces depend on the shape of the object, the speed of the object, the mass of the fluid going by the object and on two other important properties of the fluid; the viscosity, or stickiness, and the compressibility, or springiness, of the fluid. To properly model these effects, aerospace engineers use similarity parameters which are ratios of these effects to other forces present in the problem. If two experiments have the same values for the similarity parameters, then the relative importance of the forces are being correctly modeled.

Aerodynamic forces depend in a complex way on the viscosity of the fluid. As the fluid moves past the object, the molecules right next to the surface stick to the surface. The molecules just above the surface are slowed down in their collisions with the molecules sticking to the surface. These molecules in turn slow down the flow just above them. The farther one moves away from the surface, the fewer the collisions affected by the object surface. This creates a thin layer of fluid near the surface in which the velocity changes from zero at the surface to the free stream value away from the surface. Engineers call this layer the boundary layer because it occurs on the boundary of the fluid.

The details of the flow within the boundary layer are very important for many problems in aerodynamics, including wing stall, the skin friction drag on an object, and the heat transfer that occurs in high speed flight.

Types of Boundary Layer

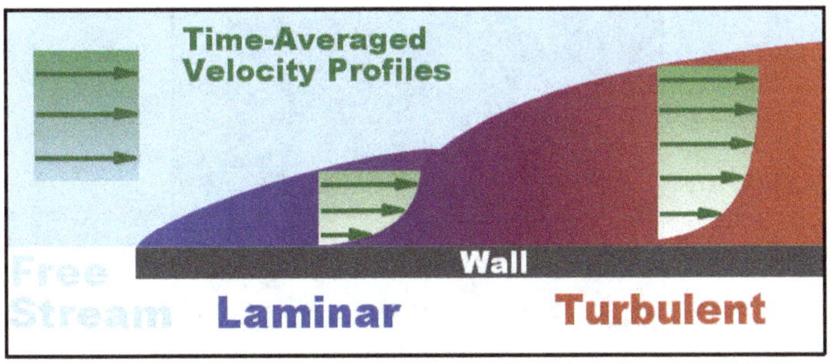

Boundary layer visualization, showing transition from laminar to turbulent condition.

Laminar boundary layers can be loosely classified according to their structure and the circumstances under which they are created. The thin shear layer which develops on an oscillating body is an example of a Stokes boundary layer, while the Blasius boundary layer refers to the well-known similarity solution near an attached flat plate held in an oncoming unidirectional flow and Falkner–Skan boundary layer, a generalization of Blasius profile. When a fluid rotates and viscous forces are balanced by the Coriolis effect (rather than convective inertia), an Ekman layer forms. In the theory of heat transfer, a thermal boundary layer occurs. A surface can have multiple types of boundary layer simultaneously.

The viscous nature of airflow reduces the local velocities on a surface and is responsible for skin friction. The layer of air over the wing's surface that is slowed down or stopped by viscosity, is the boundary layer. There are two different types of boundary layer flow: laminar and turbulent.

Laminar Boundary Layer Flow

The laminar boundary is a very smooth flow, while the turbulent boundary layer contains swirls or "eddies." The laminar flow creates less skin friction drag than the turbulent flow, but is less stable. Boundary layer flow over a wing surface begins as a smooth laminar flow. As the flow continues back from the leading edge, the laminar boundary layer increases in thickness.

Turbulent Boundary Layer Flow

At some distance back from the leading edge, the smooth laminar flow breaks down and transitions to a turbulent flow. From a drag standpoint, it is advisable to have the transition from laminar to turbulent flow as far aft on the wing as possible, or have a large amount of the wing surface within the laminar portion of the boundary layer. The low energy laminar flow, however, tends to break down more suddenly than the turbulent layer.

Aerodynamics

Ludwig Prandtl.

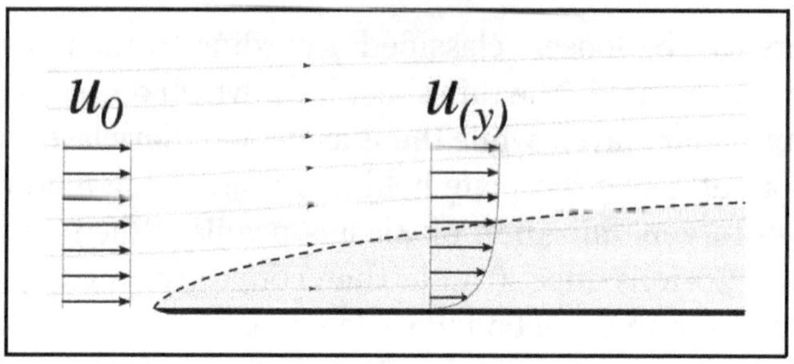

Laminar boundary layer velocity profile.

The aerodynamic boundary layer was first defined by Ludwig Prandtl in a paper presented on August 12, 1904 at the third International Congress of Mathematicians in Heidelberg, Germany. It simplifies the equations of fluid flow by dividing the flow field into two areas: one inside the boundary layer, dominated by viscosity and creating the majority of drag experienced by the boundary body; and one outside the boundary layer, where viscosity can be neglected without significant effects on the solution. This allows a closed-form solution for the flow in both areas, a significant simplification of the full Navier–Stokes equations. The majority of the heat transfer to and from a body also takes place within the boundary layer, again allowing the equations to be simplified in the flow field outside the boundary layer. The pressure distribution throughout the boundary layer in the direction normal to the surface (such as an airfoil) remains constant throughout the boundary layer, and is the same as on the surface itself.

The thickness of the velocity boundary layer is normally defined as the distance from the solid body to the point at which the viscous flow velocity is 99% of the freestream velocity (the surface velocity of an inviscid flow). Displacement thickness is an alternative definition stating that the boundary layer represents a deficit in mass flow compared to inviscid flow with slip at the wall. It is the distance by which the wall would have to be displaced in the inviscid case to give the same total mass flow as the viscous case. The no-slip condition requires the flow velocity at the surface of a solid object be zero and the fluid temperature be equal to the temperature of the surface. The flow velocity will then increase rapidly within the boundary layer, governed by the boundary layer equations.

The thermal boundary layer thickness is similarly the distance from the body at which the temperature is 99% of the freestream temperature. The ratio of the two thicknesses is governed by the Prandtl number. If the Prandtl number is 1, the two boundary layers are the same thickness. If the Prandtl number is greater than 1, the thermal boundary layer is thinner than the velocity boundary layer. If the Prandtl number is less than 1, which is the case for air at standard conditions, the thermal boundary layer is thicker than the velocity boundary layer.

In high-performance designs, such as gliders and commercial aircraft, much attention is paid to controlling the behavior of the boundary layer to minimize drag. Two effects have to be considered. First, the boundary layer adds to the effective thickness of the body, through the displacement thickness, hence increasing the pressure drag. Secondly, the shear forces at the surface of the wing create skin friction drag.

At high Reynolds numbers, typical of full-sized aircraft, it is desirable to have a laminar boundary layer. This results in a lower skin friction due to the characteristic velocity profile of laminar flow. However, the boundary layer inevitably thickens and becomes less stable as the flow develops along the body, and eventually becomes turbulent, the process known as boundary layer transition. One way of dealing with this problem is to suck the boundary layer away through a porous surface. This can reduce drag, but is usually impractical due to its mechanical complexity and the power required to move the air and dispose of it. Natural laminar flow techniques push the boundary layer transition aft by reshaping the aerofoil or fuselage so that its thickest point is more aft and less thick. This reduces the velocities in the leading part and the same Reynolds number is achieved with a greater length.

At lower Reynolds numbers, such as those seen with model aircraft, it is relatively easy to maintain laminar flow. This gives low skin friction, which is desirable. However, the same velocity profile which gives the laminar boundary layer its low skin friction also causes it to be badly affected by adverse pressure gradients. As the pressure begins to recover over the rear part of the wing chord, a laminar boundary layer will tend to separate from the surface. Such flow separation causes a large increase in the pressure drag, since it greatly increases the effective size of the wing section. In these cases, it can be advantageous to deliberately trip the boundary layer into turbulence at a point prior to the location of laminar separation, using a turbulator. The fuller velocity profile of the turbulent boundary layer allows it to sustain the adverse pressure gradient without separating. Thus, although the skin friction is increased, overall drag is decreased. This is the principle behind the dimpling on golf balls, as well as vortex generators on aircraft. Special wing sections have also been designed which tailor the pressure recovery so laminar separation is reduced or even eliminated. This represents an optimum compromise between the pressure drag from flow separation and skin friction from induced turbulence.

When using half-models in wind tunnels, a peniche is sometimes used to reduce or eliminate the effect of the boundary layer.

Boundary Layer Equations

The deduction of the boundary layer equations was one of the most important advances in fluid dynamics. Using an order of magnitude analysis, the well-known governing Navier–Stokes equations of viscous fluid flow can be greatly simplified within the boundary layer. Notably, the characteristic of the partial differential equations (PDE) becomes parabolic, rather than the elliptical form of the full Navier–Stokes equations. This greatly simplifies the solution of the equations. By making the boundary layer approximation, the flow is divided into an inviscid portion (which is easy to solve by a number of methods) and the boundary layer, which is governed by an easier to solve PDE. The continuity and Navier–Stokes equations for a two-dimensional steady incompressible flow in Cartesian coordinates are given by:

$$\frac{\partial u}{\partial x}+\frac{\partial \upsilon}{\partial y}=0$$

$$u\frac{\partial u}{\partial x}+\upsilon\frac{\partial u}{\partial y}=-\frac{1}{\rho}\frac{\partial p}{\partial x}+\nu\left(\frac{\partial^2 u}{\partial x^2}+\frac{\partial^2 u}{\partial y^2}\right)$$

$$u\frac{\partial \upsilon}{\partial x}+\upsilon\frac{\partial \upsilon}{\partial y}=-\frac{1}{\rho}\frac{\partial p}{\partial y}+\nu\left(\frac{\partial^2 \upsilon}{\partial x^2}+\frac{\partial^2 \upsilon}{\partial y^2}\right)$$

where u and υ are the velocity components, ρ is the density, p is the pressure, and ν is the kinematic viscosity of the fluid at a point.

The approximation states that, for a sufficiently high Reynolds number the flow over a surface can be divided into an outer region of inviscid flow unaffected by viscosity (the majority of the flow), and a region close to the surface where viscosity is important (the boundary layer). Let u and υ be streamwise and transverse (wall normal) velocities respectively inside the boundary layer. Using scale analysis, it can be shown that the above equations of motion reduce within the boundary layer to become.

$$u\frac{\partial u}{\partial x}+\upsilon\frac{\partial u}{\partial y}=-\frac{1}{\rho}\frac{\partial p}{\partial x}+\nu\frac{\partial^2 u}{\partial y^2}$$

$$\frac{1}{\rho}\frac{\partial p}{\partial y}=0$$

and if the fluid is incompressible (as liquids are under standard conditions):

$$\frac{\partial u}{\partial x}+\frac{\partial \upsilon}{\partial y}=0$$

The order of magnitude analysis assumes the streamwise length scale significantly larger than the transverse length scale inside the boundary layer. It follows that variations in properties in the streamwise direction are generally much lower than those in the wall normal direction. Apply this to the continuity equation shows that υ, the wall normal velocity, is small compared with u the streamwise velocity.

Since the static pressure p is independent of y, then pressure at the edge of the boundary layer is the pressure throughout the boundary layer at a given streamwise position. The external pressure may be obtained through an application of Bernoulli's equation. Let U be the fluid velocity outside the boundary layer, where u and U are both parallel. This gives upon substituting for p the following result:

$$u\frac{\partial u}{\partial x}+\upsilon\frac{\partial u}{\partial y}=U\frac{dU}{dx}+\nu\frac{\partial^2 u}{\partial y^2}$$

For a flow in which the static pressure p also does not change in the direction of the flow:

$$\frac{dp}{dx}=0$$

so U remains constant.

Therefore, the equation of motion simplifies to become:

$$u\frac{\partial u}{\partial x}+\upsilon\frac{\partial u}{\partial y}=\nu\frac{\partial^2 u}{\partial y^2}$$

These approximations are used in a variety of practical flow problems of scientific and engineering interest. The above analysis is for any instantaneous laminar or turbulent boundary layer, but is used mainly in laminar flow studies since the mean flow is also the instantaneous flow because there are no velocity fluctuations present. This simplified equation is a parabolic PDE and can be solved using a similarity solution often referred to as the Blasius boundary layer.

Prandtl's Transposition Theorem

Prandtl observed that from any solution $u(x,y,t), v(x,y,t)$ which satisfies the boundary layer equations, further solution $u^*(x,y,t), v^*(x,y,t)$, which is also satisfying the boundary layer equations, can be constructed by writing:

$$u^*(x,y,t)=u(x,y+f(x),t), \quad v^*(x,y,t)=v(x,y+f(x),t)-f'(x)u(x,y+f(x),t)$$

where $f(x)$ is arbitrary. Since the solution is not unique from mathematical perspective, to the solution can added any one of an infinite set of eigenfunctions as shown by Stewartson and Paul A. Libby.

Von Kármán Momentum Integral

Von Kármán derived the integral equation by integrating the boundary layer equation across the boundary layer in 1921. The equation is:

$$\frac{\tau_w}{\rho U^2}=\frac{1}{U^2}\frac{\partial}{\partial t}(U\delta_1)+\frac{\partial \delta_2}{\partial x}+\frac{2\delta_2+\delta_1}{U}\frac{\partial U}{\partial x}+\frac{v_w}{U}$$

where

$$\tau_w = \mu \left(\frac{\partial u}{\partial y} \right)_{y=0}, \quad v_w = v(x,0,t), \quad \delta_1 = \int_0^\infty \left(1 - \frac{u}{U} \right) dy, \quad \delta_2 = \int_0^\infty \frac{u}{U} \left(1 - \frac{u}{U} \right) dy$$

τ_w is the wall shear stress, v_w is the suction/injection velocity at the wall, δ_1 is the displacement thickness and δ_2 is the momentum thickness. Kármán-Pohlhausen Approximation is derived from this equation.

Energy Integral

The energy integral was derived by Wieghardt.

$$\frac{2\varepsilon}{\rho U^3} = \frac{1}{U} \frac{\partial}{\partial t} (\delta_1 + \delta_2) + \frac{2\delta_2}{U^2} \frac{\partial U}{\partial t} + \frac{1}{U^3} \frac{\partial}{\partial x} (U^3 \delta_3) + \frac{v_w}{U}$$

where

$$\varepsilon = \int_0^\infty \mu \left(\frac{\partial u}{\partial y} \right)^2 dy, \quad \delta_3 = \int_0^\infty \frac{u}{U} \left(1 - \frac{u^2}{U^2} \right) dy$$

ε is the energy dissipation rate due to viscosity across the boundary layer and δ_3 is the energy thickness.

Von Mises Transformation

For steady two-dimensional boundary layers, von Mises introduced a transformation which takes x and ψ (stream function) as independent variables instead of x and y and uses a dependent variable $\chi = U^2 - u^2$ instead of u. The boundary layer equation then become:

$$\frac{\partial \chi}{\partial x} = \nu \sqrt{U^2 - \chi} \frac{\partial^2 \chi}{\partial \psi^2}$$

The original variables are recovered from:

$$y = \int \sqrt{U^2 - \chi}\, d\psi, \quad u = \sqrt{U^2 - \chi}, \quad v = u \int \frac{\partial}{\partial x} \left(\frac{1}{u} \right) d\psi.$$

This transformation is later extended to compressible boundary layer by von Kármán and HS Tsien.

Crocco's Transformation

For steady two-dimensional compressible boundary layer, Luigi Crocco introduced a transformation which takes x and u as independent variables instead of x and y and uses a dependent variable $\tau = \mu \partial u / \partial y$ (shear stress) instead of u. The boundary layer equation then become:

$$\mu \rho u \frac{\partial}{\partial x} \left(\frac{1}{\tau} \right) + \frac{\partial^2 \tau}{\partial u^2} - \mu \frac{dp}{dx} \frac{\partial}{\partial u} \left(\frac{1}{\tau} \right) = 0$$

$$\text{if } \frac{dp}{dx} = 0, \quad \frac{\mu\rho}{\tau^2}\frac{\partial \tau}{\partial x} = \frac{1}{u}\frac{\partial^2 \tau}{\partial u^2}$$

The original coordinate is recovered from:

$$y = \mu \int \frac{du}{\tau}.$$

Turbulent Boundary Layers

The treatment of turbulent boundary layers is far more difficult due to the time-dependent variation of the flow properties. One of the most widely used techniques in which turbulent flows are tackled is to apply Reynolds decomposition. Here the instantaneous flow properties are decomposed into a mean and fluctuating component. Applying this technique to the boundary layer equations gives the full turbulent boundary layer equations not often given in literature:

$$\frac{\partial \overline{u}}{\partial x} + \frac{\partial \overline{v}}{\partial y} =$$

$$\overline{u}\frac{\partial \overline{u}}{\partial x} + \overline{v}\frac{\partial \overline{u}}{\partial y} = -\frac{1}{\rho}\frac{\partial \overline{p}}{\partial x} + \nu\left(\frac{\partial^2 \overline{u}}{\partial x^2} + \frac{\partial^2 \overline{u}}{\partial y^2}\right) - \frac{\partial}{\partial y}(\overline{u'v'}) - \frac{\partial}{\partial x}(\overline{u'^2})$$

$$\overline{u}\frac{\partial \overline{v}}{\partial x} + \overline{v}\frac{\partial \overline{v}}{\partial y} = -\frac{1}{\rho}\frac{\partial \overline{p}}{\partial y} + \nu\left(\frac{\partial^2 \overline{v}}{\partial x^2} + \frac{\partial^2 \overline{v}}{\partial y^2}\right) - \frac{\partial}{\partial x}(\overline{u'v'}) - \frac{\partial}{\partial y}(\overline{v'^2})$$

Using a similar order-of-magnitude analysis, the above equations can be reduced to leading order terms. By choosing length scales δ for changes in the transverse-direction, and L for changes in the streamwise-direction, with $\delta << L$, the x-momentum equation simplifies to:

$$u\frac{\partial \overline{u}}{\partial x} + \overline{v}\frac{\partial \overline{u}}{\partial y} = -\frac{1}{\rho}\frac{\partial \overline{p}}{\partial x} - \frac{\partial}{\partial y}(\overline{u'v'}).$$

This equation does not satisfy the no-slip condition at the wall. Like Prandtl did for his boundary layer equations, a new, smaller length scale must be used to allow the viscous term to become leading order in the momentum equation. By choosing $\eta << \delta$ as the y-scale, the leading order momentum equation for this "inner boundary layer" is given by:

$$0 = -\frac{1}{\rho}\frac{\partial \overline{p}}{\partial x} + \nu\frac{\partial^2 \overline{u}}{\partial y^2} - \frac{\partial}{\partial y}(\overline{u'v'}).$$

In the limit of infinite Reynolds number, the pressure gradient term can be shown to have no effect on the inner region of the turbulent boundary layer. The new "inner length scale" η is a viscous length scale, and is of order $\frac{\nu}{u_*}$, with u_* being the velocity scale of the turbulent fluctuations, in this case a friction velocity.

Unlike the laminar boundary layer equations, the presence of two regimes governed by different sets of flow scales (i.e. the inner and outer scaling) has made finding a universal similarity solution for the turbulent boundary layer difficult and controversial. To find a similarity solution that spans both regions of the flow, it is necessary to asymptotically match the solutions from both regions of the flow. Such analysis will yield either the so-called log-law or power-law.

The additional term $\overline{u'v'}$ in the turbulent boundary layer equations is known as the Reynolds shear stress and is unknown a priori. The solution of the turbulent boundary layer equations therefore necessitates the use of a turbulence model, which aims to express the Reynolds shear stress in terms of known flow variables or derivatives. The lack of accuracy and generality of such models is a major obstacle in the successful prediction of turbulent flow properties in modern fluid dynamics.

A constant stress layer exists in the near wall region. Due to the damping of the vertical velocity fluctuations near the wall, the Reynolds stress term will become negligible and we find that a linear velocity profile exists. This is only true for the very near wall region.

Compressible Flow

As an aircraft moves through the air, the air molecules near the aircraft are disturbed and move around the aircraft. If the aircraft passes at a low speed, typically less than 250 mph, it is observed that the density of the air remains constant. For higher speeds, some of the energy of the aircraft goes into compressing the air and locally changing the density of the air. This compressibility effect alters the amount of resulting force on the aircraft since the aerodynamic force depends on the air density. The effect becomes more important as speed increases. Near and beyond the speed of sound, about 330 m/s or 760 mph at sea level, small disturbances in the flow are transmitted to other locations isentropically or with constant entropy. Sharp disturbances generate shock waves that affect both the lift and drag of the aircraft, and the flow conditions downstream of the shock wave.

The Mach number is the ratio of the speed of the aircraft, or the speed of the gas, to the speed of sound in the gas. The speed of sound is equal to the speed of transmission of small, isentropic disturbances in the flow. To determine the role of the Mach number on compressibility effects. we begin with the conservation of momentum equation:

rho * V dV = - dp

where rho is the fluid density, V is the velocity, and p is the pressure. dV and dp denote differential changes in the velocity and pressure. From our derivation of the conditions for isentropic flow, we know that:

dp/p = gamma * drho/rho

dp = gamma * p / rho * drho

where gamma is the specific heat ratio. We can use the ideal equation of state to simplify the expression on the right;

$p = rho * R * T$

$dp = gamma * R * T * drho$

where R is the specific gas constant and T is the absolute temperature. We recognize that:

$gamma * R * T = a^2$

where a is equal to the speed of sound. So,

$dp = a^2 * drho$

Substituting this expression for the change of pressure into the conservation of momentum equation gives:

$rho * V\, dV = - a^2\, drho$

$- (V^2 / a^2)\, dV / V = drho / rho$

$- M^2\, dV / V = drho / rho$

where M is the Mach number.

What does this expression tell us about the role of the Mach number in compressible flows?

1. For low speed, or subsonic conditions, the Mach number is less than one, $M < 1$ and the square of the Mach number is very small. Then the left hand side of the equation is very small, and the change in density is very small. For the low subsonic conditions, compressibility can be ignored.
2. As the speed of the object approaches the speed of sound, the flight Mach number is nearly equal to one, $M = 1$, and the flow is said to be transonic. If the Mach number is near one, the square of the Mach number is also nearly equal to one. For transonic flows, the change in density is nearly equal to the change in velocity, and compressibility effects can not be ignored.
3. As the speed increases beyond the speed of sound, the flight Mach number is greater than one $M > 1$ and the flow is said to be supersonic or hypersonic. For supersonic and hypersonic flows, the density changes faster than the velocity changes by a factor equl to the square of the Mach number. Compressibility effects become more important with higher Mach numbers.

Introductory Concepts

There are several important assumptions involved in the underlying theory of compressible flow. All fluids are composed of molecules, but tracking a huge number of individual molecules in a flow (for example at atmospheric pressure) is unnecessary. Instead, the continuum assumption allows us to consider a flowing gas as a continuous substance except at low densities. This assumption provides a huge simplification which is accurate for most gas-dynamic problems. Only in the low-density realm of rarefied gas dynamics does the motion of individual molecules become important.

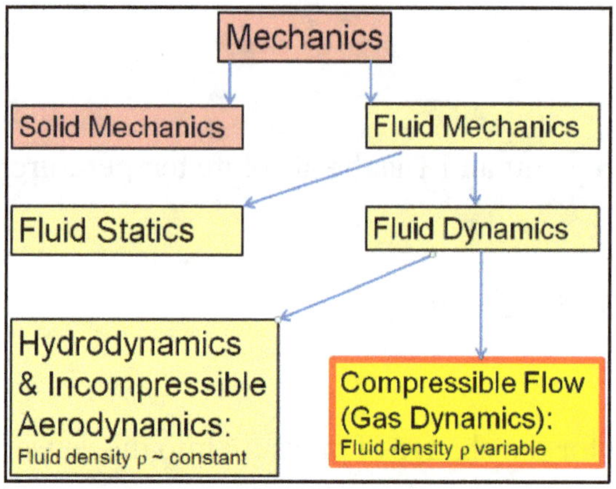

Breakdown of fluid mechanics chart.

A related assumption is the no-slip condition where the flow velocity at a solid surface is presumed equal to the velocity of the surface itself, which is a direct consequence of assuming continuum flow. The no-slip condition implies that the flow is viscous, and as a result a boundary layer forms on bodies traveling through the air at high speeds, much as it does in low-speed flow.

Most problems in incompressible flow involve only two unknowns: pressure and velocity, which are typically found by solving the two equations that describe conservation of mass and of linear momentum, with the fluid density presumed constant. In compressible flow, however, the gas density and temperature also become variables. This requires two more equations in order to solve compressible-flow problems: an equation of state for the gas and a conservation of energy equation. For the majority of gas-dynamic problems, the simple Ideal gas law is the appropriate state equation.

Fluid dynamics problems have two overall types of references frames, called Lagrangian and Eulerian. The Lagrangian approach follows a fluid mass of fixed identity as it moves through a flowfield. The Eulerian reference frame, in contrast, does not move with the fluid. Rather it is a fixed frame or control volume that fluid flows through. The Eulerian frame is most useful in a majority of compressible flow problems, but requires that the equations of motion be written in a compatible format.

Finally, although space is known to have 3 dimensions, an important simplification can be had in describing gas dynamics mathematically if only one spatial dimension is of primary importance, hence 1-dimensional flow is assumed. This works well in duct, nozzle, and diffuser flows where the flow properties change mainly in the flow direction rather than perpendicular to the flow. However, an important class of compressible flows, including the external flow over bodies traveling at high speed, requires at least a 2-dimensional treatment. When all 3 spatial dimensions and perhaps the time dimension as well are important, we often resort to computerized solutions of the governing equations.

One-dimensional Flow

One-dimensional (1-D) flow refers to the flow of gas through a duct or channel in which the flow parameters are assumed to change significantly along only one spatial dimension, namely, the duct length. In analysing the 1-D channel flow, a number of assumptions are made:

- Ratio of duct length to width (L/D) is ≤ about 5 (in order to neglect friction and heat transfer),
- Steady vs. Unsteady Flow,

- Flow is isentropic (i.e. a reversible adiabatic process),
- Ideal gas law (i.e. $P=\rho RT$).

Converging-diverging Laval Nozzles

As the speed of a flow accelerates from the subsonic to the supersonic regime, the physics of nozzle and diffuser flows is altered. Using the conservation laws of fluid dynamics and thermodynamics, the following relationship for channel flow is developed (combined mass and momentum conservation):

$$dP(1-M^2)=\rho V^2\left(\frac{dA}{A}\right),$$

where dP is the differential change in pressure, M is the Mach number, ρ is the density of the gas, V is the velocity of the flow, A is the area of the duct, and dA is the change in area of the duct. This equation states that, for subsonic flow, a converging duct ($dA<0$) increases the velocity of the flow and a diverging duct ($dA>0$) decreases velocity of the flow. For supersonic flow, the opposite occurs due to the change of sign of ($1-M^2$). A converging duct ($dA<0$) now decreases the velocity of the flow and a diverging duct ($dA>0$) increases the velocity of the flow. At Mach = 1, a special case occurs in which the duct area must be either a maximum or minimum. For practical purposes, only a minimum area can accelerate flows to Mach 1 and beyond. See table of sub-supersonic diffusers and nozzles.

	Subsonic M<1	Supersonic M>1
Converging Duct 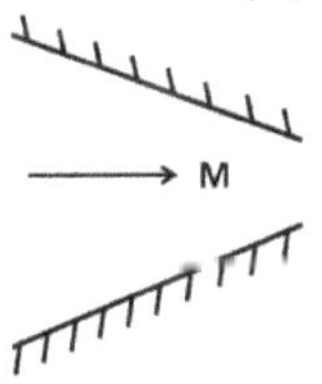	As P decreases, then v increases Subsonic nozzle	As P increases, then v decreases Supersonic diffuser
Diverging Duct 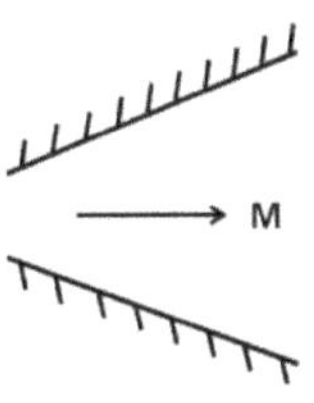	As P increases, then v decreases Subsonic diffuser	As P decreases, then v increases Supersonic nozzle

Table showing the reversal in the physics of nozzles and diffusers with changing Mach numbers.

Therefore, to accelerate a flow to Mach 1, a nozzle must be designed to converge to a minimum cross-sectional area and then expand. This type of nozzle – the converging-diverging nozzle – is called a de Laval nozzle after Gustaf de Laval, who invented it. As subsonic flow enters the converging duct and the area decreases, the flow accelerates. Upon reaching the minimum area of the duct, also known as the throat of the nozzle, the flow can reach Mach 1. If the speed of the flow is to continue to increase, its density must decrease in order to obey conservation of mass. To achieve this decrease in density, the flow must expand, and to do so, the flow must pass through a diverging duct.

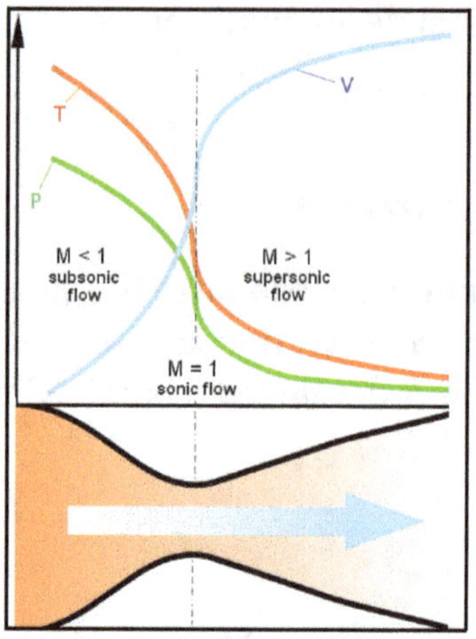

Nozzle de Laval diagram.

Maximum Achievable Velocity of a Gas

Ultimately, because of the energy conservation law, a gas is limited to a certain maximum velocity based on its energy content. The maximum velocity, V_{max}, that a gas can attain is:

$$V_{max} = \sqrt{2c_p T_t}$$

where c_p is the specific heat of the gas and T_t is the stagnation temperature of the flow.

Isentropic Flow Mach Number Relationships

Using conservations laws and thermodynamics, a number of relationships of the form

$$\frac{property_1}{property_2} = f(M,\gamma)$$

can be obtained, where M is the Mach number and γ is the ratio of specific heats (1.4 for air). See table of isentropic flow Mach number relationships.

Isentropic Flow Mach Number Relationships:	
Stagnation to Static Temperature	$\frac{T_t}{T} = \left(1 + \frac{\gamma - 1}{2} M^2\right)$
Stagnation to Static Pressure	$\frac{P_t}{P} = \left(1 + \frac{\gamma - 1}{2} M^2\right)^{\frac{\gamma}{\gamma-1}}$
Mass Flow Rate	$\dot{m} = \frac{P_t}{\sqrt{RT_t}} A\sqrt{\gamma}\, M \left(1 + \frac{\gamma - 1}{2} M^2\right)^{\frac{\gamma+1}{2-2\gamma}}$
Stagnation to Static Density	$\frac{\rho_t}{\rho} = \left(1 + \frac{\gamma - 1}{2} M^2\right)^{\frac{1}{\gamma-1}}$
Arbitrary Area to Choked Area in Duct	$\frac{A}{A*} = \frac{1}{M}\left(\frac{\frac{\gamma+1}{2}}{1 + \frac{\gamma-1}{2} M^2}\right)^{\frac{\gamma+1}{2-2\gamma}}$

Isentropic flow relationship table. Equations to relate the field properties in isentropic flow.

Achieving Supersonic Flow

In order for a flow to become supersonic, it must pass through a duct with a minimum area, or sonic throat. Additionally, an overall pressure ratio, P_b/P_t, of approximately 2 is needed to attain Mach 1. Once it has reached Mach 1, the flow at the throat is said to be "choked." Because changes downstream can only move upstream at sonic speed, the mass flow through the nozzle cannot be affected by changes in downstream conditions after the flow is choked.

Non-isentropic 1D Channel Flow of a Gas - Normal Shock Waves

Normal shock waves are shock waves that are perpendicular to the local flow direction. These shock waves occur when pressure waves build up and coalesce into an extremely thin shock-wave that converts useful energy into heat. The waves thus overtake and reinforce one another, forming a finite shock wave from an infinite series of infinitesimal sound waves. Because a loss of energy occurs over the thin shock wave, the shock is considered non-isentropic and enthalpy increases across the shock. When analysing a normal shock wave, one-dimensional, steady, and adiabatic (stagnation temperature does not change across the shock wave) flow of a perfect gas is assumed.

Rankine-Hugoniot Equations:	
Temperature	$\frac{T_2}{T_1} = \frac{\left(1+\frac{\gamma-1}{2}M_1^2\right)\left(\frac{2\gamma}{\gamma-1}M_1^2-1\right)}{M_1^2\left(\frac{2\gamma}{\gamma-1}+\frac{\gamma-1}{2}\right)}$
Static Pressure	$\frac{P_2}{P_1} = \frac{2\gamma M_1^2}{\gamma+1} - \frac{\gamma-1}{\gamma+1}$
Density and Velocity	$\frac{\rho_2}{\rho_1} = \frac{v_2}{v_1} = \frac{(\gamma+1)M_1^2}{(\gamma-1)M_1^2+2}$

The Rankine-Hugoniot equations relate conditions before and after a normal shock wave.

Normal shock waves can occur in two reference frames: the standing normal shock and the moving shock. The flow before a normal shock wave must be supersonic, and the flow after a normal shock must be subsonic. The Rankine-Hugoniot equations are used to solve for the flow conditions.

Two-dimensional Flow

Although one-dimensional flow can be directly analysed, it is merely a specialized case of two-dimensional flow. It follows that one of the defining phenomena of one-dimensional flow, a normal shock, is likewise only a special case of a larger class of oblique shocks. Further, the name "normal" is with respect to geometry rather than frequency of occurrence. Oblique shocks are much more common in applications such as: aircraft inlet design, objects in supersonic flight, and (at a more fundamental level) supersonic nozzles and diffusers. Depending on the flow conditions, an oblique shock can either be attached to the flow or detached from the flow in the form of a bow shock.

Attached shock wave shown on a X-15 Model in a supersonic wind tunnel.

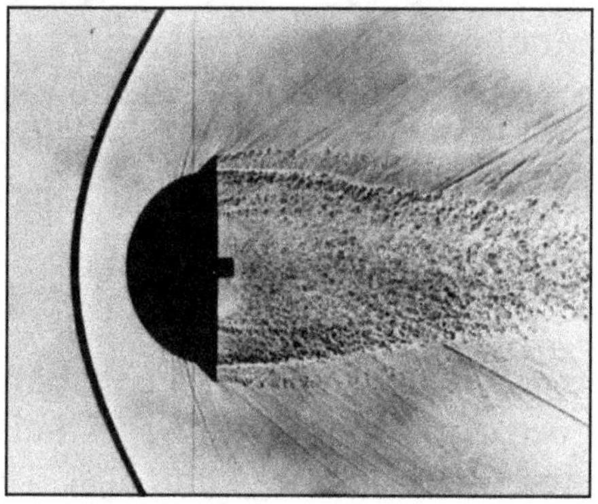

Bowshock example for a blunt body.

Oblique Shock Waves

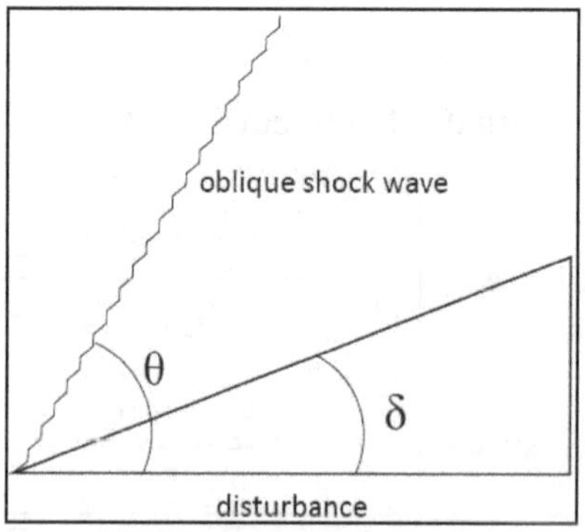

Diagram of obstruction.

Oblique shock waves are similar to normal shock waves, but they occur at angles less than 90° with the direction of flow. When a disturbance is introduced to the flow at a nonzero angle (δ), the flow must respond to the changing boundary conditions. Thus an oblique shock is formed, resulting in a change in the direction of the flow.

Shock Polar Diagram

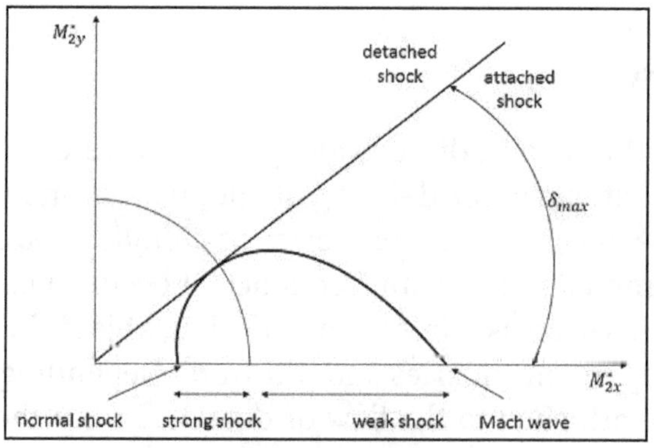

Shock polar diagram.

Based on the level of flow deflection (δ), oblique shocks are characterized as either strong or weak. Strong shocks are characterized by larger deflection and more entropy loss across the shock, with weak shocks as the opposite. In order to gain cursory insight into the differences in these shocks, a shock polar diagram can be used. With the static temperature after the shock, T*, known the speed of sound after the shock is defined as,

$$A^* = \sqrt{\gamma R T^*}$$

with R as the gas constant and γ as the specific heat ratio. The Mach number can be broken into Cartesian coordinates:

$$M_{2x}^* = \frac{V_x}{a^*}$$

$$M_{2y}^* = \frac{V_y}{a^*}$$

with V_x and V_y as the x and y-components of the fluid velocity V. With the Mach number before the shock given, a locus of conditions can be specified. At some δ_{max} the flow transitions from a strong to weak oblique shock. With δ = 0°, a normal shock is produced at the limit of the strong oblique shock and the Mach wave is produced at the limit of the weak shock wave.

Oblique Shock Reflection

Due to the inclination of the shock, after an oblique shock is created, it can interact with a boundary in three different manners.

Solid Boundary

Incoming flow is first turned by angle δ with respect to the flow. This shockwave is reflected off the solid boundary, and the flow is turned by – δ to again be parallel with the boundary. It is important to note that each progressive shock wave is weaker and the wave angle is increased.

Irregular Reflection

An irregular reflection is much like the case described above, with the caveat that δ is larger than the maximum allowable turning angle. Thus a detached shock is formed and a more complicated reflection occurs.

Prandtl-meyer Fans

Prandtl-Meyer fans can be expressed as both compression and expansion fans. Prandtl-Meyer fans also cross a boundary layer (i.e. flowing and solid) which reacts in different changes as well. When a shock wave hits a solid surface the resulting fan returns as one from the opposite family while when one hits a free boundary the fan returns as a fan of opposite type.

Prandtl-meyer Expansion Fans

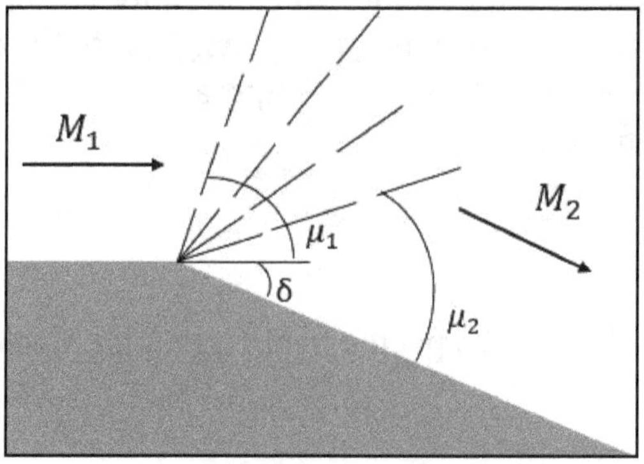

Prandtl-Meyer expansion fan diagram.

To this point, the only flow phenomena that have been discussed are shock waves, which slow the flow and increase its entropy. It is possible to accelerate supersonic flow in what has been termed a Prandtl-Meyer expansion fan, after Ludwig Prandtl and Theodore Meyer.

As opposed to the flow encountering an inclined obstruction and forming an oblique shock, the flow expands around a convex corner and forms an expansion fan through a series of isentropic Mach waves. The expansion "fan" is composed of Mach waves that span from the initial Mach angle to the final Mach angle. Flow can expand around either a sharp or rounded corner equally, as the increase in Mach number is proportional to only the convex angle of the passage (δ). The expansion corner that produces the Prandtl-Meyer fan can be sharp (as illustrated in the figure) or rounded. If the total turning angle is the same, then the P-M flow solution is also the same.

The Prandtl-Meyer expansion can be seen as the physical explanation of the operation of the Laval nozzle. The contour of the nozzle creates a smooth and continuous series of Prandtl-Meyer expansion waves.

Prandtl-meyer Compression Fans

A Prandtl-Meyer compression is the opposite phenomenon to a Prandtl-Meyer expansion. If the flow is gradually turned through an angle of δ, a compression fan can be formed. This fan is a series of Mach waves that eventually coalesce into an oblique shock. Because the flow is defined by an isentropic region (flow that travels through the fan) and an anisentropic region (flow that travels through the oblique shock), a slip line results between the two flow regions.

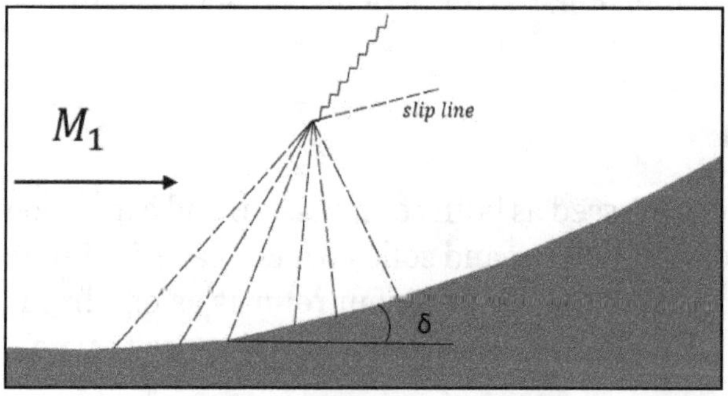

Basic PM compression diagram.

Applications

Supersonic Wind Tunnels

Supersonic wind tunnels are used for testing and research in supersonic flows, approximately over the Mach number range of 1.2 to 5. The operating principle behind the wind tunnel is that a large pressure difference is maintained upstream to downstream, driving the flow.

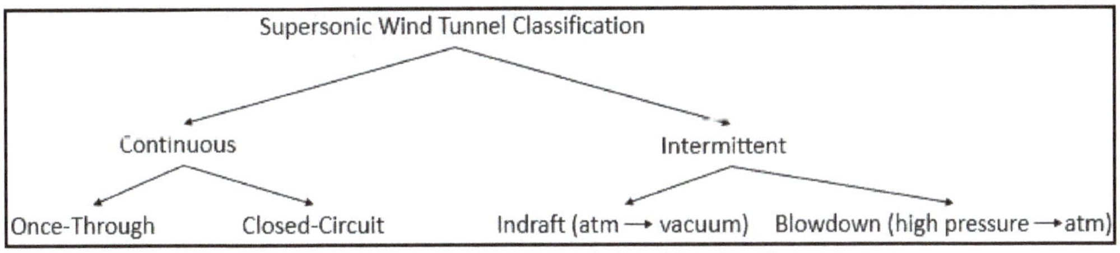

Supersonic wind tunnel classification list.

Wind tunnels can be divided into two categories: continuous-operating and intermittent-operating wind tunnels. Continuous operating supersonic wind tunnels require an independent electrical power source that drastically increases with the size of the test section. Intermittent supersonic wind tunnels are less expensive in that they store electrical energy over an extended period of time, then discharge the energy over a series of brief tests. The difference between these two is analogous to the comparison between a battery and a capacitor.

Blowdown type supersonic wind tunnels offer high Reynolds number, a small storage tank, and readily available dry air. However, they cause a high pressure hazard, result in difficulty holding a constant stagnation pressure, and are noisy during operation.

Indraft supersonic wind tunnels are not associated with a pressure hazard, allow a constant stagnation pressure, and are relatively quiet. Unfortunately, they have a limited range for the Reynolds number of the flow and require a large vacuum tank. There is no dispute that knowledge is gained through research and testing in supersonic wind tunnels; however, the facilities often require vast amounts of power to maintain the large pressure ratios needed for testing conditions. For example, Arnold Engineering Development Complex has the largest supersonic wind tunnel in the world and requires the power required to light a small city for operation. For this reason, large wind tunnels are becoming less common at universities.

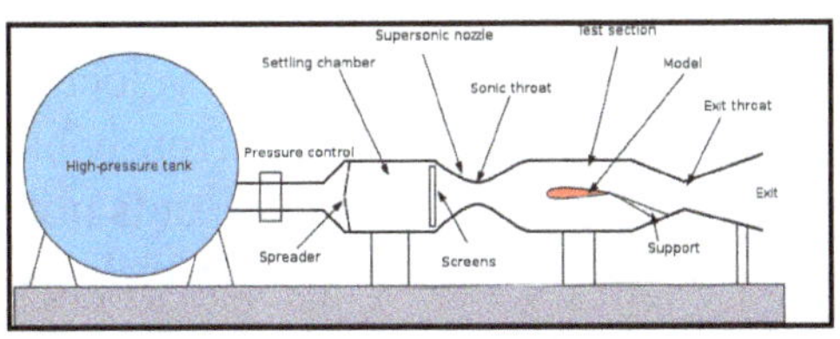

Blowdown supersonic wind tunnel schematic.

Langley indraft supersonic wind tunnel vacuum sphere.

Supersonic Aircraft Inlets

Perhaps the most common application for oblique shocks is in high-speed aircraft inlets. The purpose of the inlet is to slow incoming supersonic flow to the subsonic regime before it enters the turbojet engine, with the caveat of minimizing losses across the shock. Knowledge of normal and oblique shocks suggests that this be accomplished with a series of weakening oblique shocks followed by a very weak normal shock, usually less than M = 1.4. This may sound relatively straightforward, but there is one rather large issue to be dealt with when designing a supersonic aircraft inlet: acceleration. Between taking off, manoeuvring, and cruising, an aircraft travels at a range of Mach numbers. In order to ensure efficient flight, the aircraft intake must be capable of variable geometry. If it is not, the shock waves will not reflect properly through the inlet and negatively affect performance.

McDonnell Douglas F-15C.

Concorde.

Lockheed SR-71 Blackbird.

Although variable geometry is a universally recognized approach to improve aircraft efficiency and performance over a range of Mach numbers, there is no one method to achieve variable geometry. The F-15 Eagle employs wedge inlets with adjustable flaps to control the flow. For subsonic flow, the flaps are completely closed and for supersonic flow, the flaps are open. The Concorde employed an external-compression inlet, using a series of oblique shocks followed by a normal shock to slow the flow sufficiently for the turbojet engine. Perhaps the most recognizable supersonic aircraft, the SR-71, used a hydraulically actuated cone to reduce the speed of the supersonic flow through the aircraft inlet.

Transonic

As an aircraft moves through the air, the air molecules near the aircraft are disturbed and move around the aircraft. Exactly how the air re-acts to the aircraft depends upon the ratio of the speed of the aircraft to the speed of sound through the air. Because of the importance of this speed ratio, aerodynamicists have designated it with a special parameter called the Mach number in honor of Ernst Mach, a late 19th century physicist who studied gas dynamics.

In aeronautics, transonic refers to the condition of flight in which a range of velocities of airflow exist surrounding and flowing past an air vehicle or an airfoil that are concurrently below, at, and above the speed of sound in the range of Mach 0.8 to 1.2. This condition depends not only on the travel speed of the craft, but also on the temperature of the airflow in the vehicle's local environment. It is formally defined as the range of speeds between the critical Mach number, when some parts of the airflow over an air vehicle or airfoil are supersonic, and a higher speed, typically near Mach 1.2, when the vast majority of the airflow is supersonic. Between these speeds some of the airflow is supersonic, but a significant fraction is not.

The first powered aircraft to explore this regime were the high performance fighters of World War II. These aircraft seemed to encounter a sound barrier at which drag was increasing faster than thrust. There was speculation in the mid-1940's that manned flight was not possible at speeds faster than the speed of sound, even though the muzzle velocity of rifle bullets is supersonic. Of course, the flight of the X-1A in 1947 proved that people could fly faster than sound and, until the recent retirement of the Concorde, any person with enough money can fly supersonic. As mentioned above, even though modern airliners typically fly at about M = .85, the flow over the wings is transonic or supersonic. Drag increases dramatically as an aircraft approaches Mach 1, so airliners use high thrust gas turbine propulsion systems. On the slide we show a DC-8 airliner which is powered by four turbofan engines. The wings of airliners are typically swept in planform to reduce the transonic drag. For Mach numbers less than 2.0, the frictional heating of the airframe is low enough that light weight aluminum is used for the structure.

Most modern jet powered aircraft are engineered to operate at transonic air speeds. Transonic airspeeds see a rapid increase of drag from about Mach 0.8, and it is the fuel costs of that drag that typically limits the airspeed. Attempts to reduce wave drag can be seen on all high-speed aircraft; most notable is the use of swept wings, but another common form is a wasp-waist fuselage as a side effect of the Whitcomb area rule.

Transonic cloud forms around shock wave as aircraft passes through the sound barrier.

Severe instability can occur at transonic speeds. Shock waves move through the air at the speed of sound. When an object such as an aircraft also moves at the speed of sound, these shock waves build up in front of it to form a single, very large shock wave. During transonic flight, the plane must pass through this large shock wave, as well as contend with the instability caused by air moving faster than sound over parts of the wing and slower in other parts.

Transonic speeds can also occur at the tips of rotor blades of helicopters and aircraft. However, as this puts severe, unequal stresses on the rotor blade and may lead to dangerous accidents if it occurs, it is avoided. Blade tip speed is one of the limiting factors to the size of rotors, and also to the forward speeds of helicopters (as this speed is added to the forward-sweeping (leading) side of the rotor, thus possibly causing localized transonics).

Condensation Clouds

At transonic speeds supersonic expansion fans form intense low-pressure, low-temperature areas at various points around an aircraft. If the temperature drops below the dew point a visible cloud will form. These clouds remain with the aircraft as it travels. It is not necessary for the aircraft as a whole to reach supersonic speeds for these clouds to form. Typically, the tail of the aircraft will reach supersonic flight while the bow of the aircraft is still in subsonic flight. A bubble of supersonic expansion fans terminating by a wake shockwave surround the tail. As the aircraft continues to accelerate, the supersonic expansion fans will intensify and the wake shockwave will grow in size until infinity is reached, at which point the bow shockwave forms. This is Mach one and the Prandtl–Glauert singularity.

Mach Number

Mach number, in fluid mechanics, ratio of the velocity of a fluid to the velocity of sound in that fluid, named after Ernst Mach (1838–1916), an Austrian physicist and philosopher. In the case of an object moving through a fluid, such as an aircraft in flight, the Mach number is equal to the velocity of the object relative to the fluid divided by the velocity of sound in that fluid. Mach numbers less than one indicate subsonic flow; those greater than one, supersonic flow. Fluid flow, in addition, is classified as compressible or incompressible on the basis of the Mach number. For example, gas flowing with a Mach number of less than three-tenths may be considered incompressible, or of constant density, an approximation that greatly simplifies the analysis of its behaviour. For Mach numbers greater than one, shock wave patterns develop on the moving body because of compression of the surrounding fluid. Streamlining alleviates shock wave effects.

The Mach Number is a dimensionless value useful for analyzing fluid flow dynamics problems where compressibility is a significant factor.

The Mach Number can be expressed as:

$$M = v / c$$

where

M = Mach number

v = fluid flow speed (m/s, ft/s)

c = speed of sound (m/s, ft/s)

Alternatively the Mach Number can be expressed with the density and the bulk modulus for elasticity as:

$$M = v\,(\rho/E)^{1/2}$$

where

ρ = density of fluid (kg/m^3, lb/ft^3)

E = bulk modulus elasticity (N/m^2 (Pa), lbf/in^2 (psi))

The bulk modulus elasticity has the dimension pressure and is commonly used to characterize the fluid compressibility.

The square of the Mach number is the Cauchy Number.

$$M^2 = C$$

where

C = Cauchy Number

Mach number is useful because the fluid behaves in a similar manner at a given Mach number, regardless of other variables. As modeled in the International Standard Atmosphere, dry air at mean sea level, standard temperature of 15 °C (59 °F), the speed of sound is 340.3 meters per second (1,116.5 ft/s). The speed of sound is not a constant; in a gas, it increases as the absolute temperature increases, and since atmospheric temperature generally decreases with increasing altitude between sea level and 11,000 meters (36,089 ft), the speed of sound also decreases. For example, the standard atmosphere model lapses temperature to −56.5 °C (−69.7 °F) at 11,000 meters (36,089 ft) altitude, with a corresponding speed of sound (Mach 1) of 295.0 meters per second (967.8 ft/s), 86.7% of the sea level value.

Classification of Mach Regimes

While the terms *subsonic* and *supersonic*, in the purest sense, refer to speeds below and above the local speed of sound respectively, aerodynamicists often use the same terms to talk about particular ranges of Mach values. This occurs because of the presence of a *transonic regime* around M = 1 where approximations of the Navier-Stokes equations used for subsonic design actually no longer apply; the simplest explanation is that the flow locally begins to exceed M = 1 even though the freestream Mach number is below this value.

Meanwhile, the *supersonic regime* is usually used to talk about the set of Mach numbers for which linearised theory may be used, where for example the (air) flow is not chemically reacting, and where heat-transfer between air and vehicle may be reasonably neglected in calculations.

In the following table, the *regimes* or *ranges of Mach values* are referred to, and not the *pure* meanings of the words *subsonic* and *supersonic*.

Generally, NASA defines *high* hypersonic as any Mach number from 10 to 25, and re-entry speeds as anything greater than Mach 25. Aircraft operating in this regime include the Space Shuttle and various space planes in development.

Regime	Speed					General plane characteristics
	(Mach)	(knots)	(mph)	(km/h)	(m/s)	
Subsonic	<0.8	<530	<609	<980	<273	Most often propeller-driven and commercial turbofan aircraft with high aspect-ratio (slender) wings, and rounded features like the nose and leading edges. The subsonic speed range is that range of speeds within all of the airflow over an aircraft is less than Mach 1. The critical Mach number (Mcrit) is lowest free stream mach number at which airflow over any part of the aircraft first reaches Mach 1. So the subsonic speed range includes all speeds that are less than Mcrit.

Transonic	0.8–1.3	530–794	609–914	980 –1,470	273 –409	Transonic aircraft nearly always have swept wings, causing the delay of drag-divergence, and often feature a design that adheres to the principles of the Whitcomb Area rule. The transonic speed range is that range of speeds within which the airflow over different parts of an aircraft is between subsonic and supersonic. So the regime of flight from Mcrit up to Mach 1.3 is called the transonic range.
Supersonic	1.3–5.0	794-3,308	915-3,806	1,470 –6,126	410 –1,702	The supersonic speed range is that range of speeds within which all of the airflow over an aircraft is supersonic (more than Mach 1). But airflow meeting the leading edges is initially decelerated, so the free stream speed must be slightly greater than Mach 1 to ensure that all of the flow over the aircraft is supersonic. It is commonly accepted that the supersonic speed range starts at a free stream speed greater than Mach 1.3. Aircraft designed to fly at supersonic speeds show large differences in their aerodynamic design because of the radical differences in the behaviour of flows above Mach 1. Sharp edges, thin aerofoil-sections, and all-moving tailplane/ canards are common. Modern combat aircraft must compromise in order to maintain low-speed handling; "true" supersonic designs include the F-104 Starfighter, SR-71 Blackbird and BAC/Aérospatiale Concorde.
Hypersonic	5.0 –10.0	3,308 –6,615	3,806 –7,680	6,126 –12,251	1,702 –3,403	The X-15, at Mach 6.72 is one of the fastest manned aircraft. Also, cooled nickel-titanium skin; highly integrated (due to domination of interference effects: non-linear behaviour means that superposition of results for separate components is invalid), small wings, such as those on the Mach 5 X-51A Waverider.
High-hypersonic	10.0 –25.0	6,615 –16,537	7,680 –19,031	12,251 –30,626	3,403 –8,508	The NASA X-43, at Mach 9.6 is one of the fastest aircraft. Thermal control becomes a dominant design consideration. Structure must either be designed to operate hot, or be protected by special silicate tiles or similar. Chemically reacting flow can also cause corrosion of the vehicle's skin, with free-atomic oxygen featuring in very high-speed flows. Hypersonic designs are often forced into blunt configurations because of the aerodynamic heating rising with a reduced radius of curvature.
Re-entry speeds	>25.0	>16,537	>19,031	>30,626	>8,508	Ablative heat shield; small or no wings; blunt shape

High-speed Flow Around Objects

Flight can be roughly classified in six categories:

Regime	Subsonic	Transonic	Sonic	Supersonic	Hypersonic	Hypervelocity
Mach	<0.8	0.8–1.2	1.0	1.2–5.0	5.0–10.0	>10.0

For comparison: the required speed for low Earth orbit is approximately 7.5 km/s = Mach 25.4 in air at high altitudes.

At transonic speeds, the flow field around the object includes both sub- and supersonic parts. The transonic period begins when first zones of M > 1 flow appear around the object. In case of an airfoil (such as an aircraft's wing), this typically happens above the wing. Supersonic flow can decelerate back to subsonic only in a normal shock; this typically happens before the trailing edge.

As the speed increases, the zone of M > 1 flow increases towards both leading and trailing edges. As M = 1 is reached and passed, the normal shock reaches the trailing edge and becomes a weak oblique shock: the flow decelerates over the shock, but remains supersonic. A normal shock is created ahead of the object, and the only subsonic zone in the flow field is a small area around the object's leading edge.

When an aircraft exceeds Mach 1 (i.e. the sound barrier), a large pressure difference is created just in front of the aircraft. This abrupt pressure difference, called a shock wave, spreads backward and outward from the aircraft in a cone shape (a so-called Mach cone). It is this shock wave that causes the sonic boom heard as a fast moving aircraft travels overhead. A person inside the aircraft will not hear this. The higher the speed, the more narrow the cone; at just over M = 1 it is hardly a cone at all, but closer to a slightly concave plane.

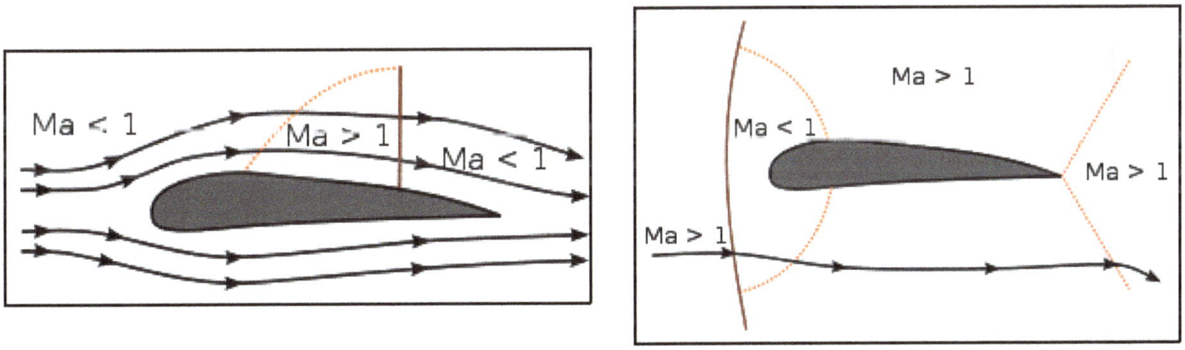

(a) (b)

Mach number in transonic airflow around an airfoil; M < 1 (a) and M > 1 (b).

At fully supersonic speed, the shock wave starts to take its cone shape and flow is either completely supersonic, or (in case of a blunt object), only a very small subsonic flow area remains between the object's nose and the shock wave it creates ahead of itself. (In the case of a sharp object, there is no air between the nose and the shock wave: the shock wave starts from the nose).

As the Mach number increases, so does the strength of the shock wave and the Mach cone becomes increasingly narrow. As the fluid flow crosses the shock wave, its speed is reduced and temperature, pressure, and density increase. The stronger the shock, the greater the changes. At high enough Mach numbers the temperature increases so much over the shock that ionization and dissociation of gas molecules behind the shock wave begin. Such flows are called hypersonic.

It is clear that any object traveling at hypersonic speeds will likewise be exposed to the same extreme temperatures as the gas behind the nose shock wave, and hence choice of heat-resistant materials becomes important.

High-speed Flow in a Channel

As a flow in a channel becomes supersonic, one significant change takes place. The conservation of mass flow rate leads one to expect that contracting the flow channel would increase the flow speed (i.e. making the channel narrower results in faster air flow) and at subsonic speeds this holds true. However, once the flow becomes supersonic, the relationship of flow area and speed is reversed: expanding the channel actually increases the speed.

The obvious result is that in order to accelerate a flow to supersonic, one needs a convergent-divergent nozzle, where the converging section accelerates the flow to sonic speeds, and the diverging section continues the acceleration. Such nozzles are called de Laval nozzles and in extreme cases they are able to reach hypersonic speeds (Mach 13 (15,926 km/h; 9,896 mph) at 20 °C).

An aircraft Machmeter or electronic flight information system (EFIS) can display Mach number derived from stagnation pressure (pitot tube) and static pressure.

Calculation

The Mach number at which an aircraft is flying can be calculated by:

$$M = \frac{u}{c}$$

where:

M is the Mach number

u is velocity of the moving aircraft and

c is the speed of sound at the given altitude

Note that the dynamic pressure can be found as:

$$q = \frac{\gamma}{2} p M^2$$

Assuming air to be an ideal gas, the formula to compute Mach number in a subsonic compressible flow is derived from Bernoulli's equation for M < 1:

$$M = \sqrt{\frac{2}{\gamma - 1}\left[\left(\frac{q_c}{p} + 1\right)^{\frac{\gamma - 1}{\gamma}} - 1\right]}$$

where:

q_c is impact pressure (dynamic pressure) and

p is static pressure

γ is the ratio of specific heat of a gas at a constant pressure to heat at a constant volume (1.4 for air).

The formula to compute Mach number in a supersonic compressible flow is derived from the Rayleigh supersonic pitot equation:

$$\frac{p_t}{p} = \left[\frac{\gamma+1}{2}M^2\right]^{\frac{\gamma}{\gamma-1}} \cdot \left[\frac{\gamma+1}{1-\gamma+2\gamma M^2}\right]^{\frac{1}{\gamma-1}}$$

Calculating Mach Number from Pitot Tube Pressure

Mach number is a function of temperature and true airspeed. Aircraft flight instruments, however, operate using pressure differential to compute Mach number, not temperature.

Assuming air to be an ideal gas, the formula to compute Mach number in a subsonic compressible flow is found from Bernoulli's equation for M < 1:

$$= \sqrt{5\left[\left(\frac{}{}+1\right)^{-} -1\right]}$$

The formula to compute Mach number in a supersonic compressible flow can be found from the Rayleigh supersonic pitot equation (above) using parameters for air:

$$M \approx 0.88128485\sqrt{\left(\frac{q_c}{p}+1\right)\left(1-\frac{1}{7M^2}\right)^{2.5}}$$

where:

q_c is the dynamic pressure measured behind a normal shock.

As can be seen, M appears on both sides of the equation, and for practical purposes a root-finding algorithm must be used for a numerical solution. It is first determined whether M is indeed greater than 1.0 by calculating M from the subsonic equation. If M is greater than 1.0 at that point, then the value of M from the subsonic equation is used as the initial condition for fixed point iteration of the supersonic equation, which usually converges in just a few iterations. Alternatively, Newton's method can also be used.

Supersonic Speed

The term supersonic is used to define a speed that exceeds the speed of sound—a speed that is referred to as Mach 1. However, supersonic airflow is decidedly different from subsonic flow, leading to some debate about the definition of "supersonic." Thus, another definition states that an aircraft (or object) is supersonic only if the airflow over the entire aircraft (or object) is supersonic, which occurs around Mach 1.2 on typical designs. The range Mach 0.75 to 1.2 is considered transonic, when only some parts of the air around the aircraft (or object) reach supersonic speeds. Speeds greater than 5 times the speed of sound are often referred to as hypersonic. Chuck Yeager was the first man to achieve supersonic flight.

Value of Mach 1

At a typical temperature, such as 21 °C (70 °F), the value of Mach 1 is approximately 344 meters per second (m/s) (1,129 ft/s, 770 mph or 1,238 km/h). Nearly everything about the way an aircraft flies changes dramatically as an aircraft accelerates to supersonic speeds.

Sound Waves in Gases, Liquids, and Solids

Sounds are traveling vibrations (pressure waves) in an elastic medium. In gases, sound travels longitudinally at different speeds, mostly depending on the molecular mass and temperature of the gas; (pressure has little effect). Because air temperature and composition vary significantly with altitude, Mach numbers for aircraft can change without variation of airspeed. In water at room temperature, supersonic can be considered as any speed greater than 1,440 m/s (4,724 ft/s). In solids, sound waves can be longitudinal or transverse and have even higher velocities.

Supersonic fracture is crack motion faster than the speed of sound in a brittle material.

Breaking the Sound Barrier

In aerodynamics, the "sound barrier" usually refers to the point at which an aircraft moves from transonic to supersonic speed. The term came into use during World War II, when a number of aircraft started to encounter the effects of compressibility, a grab-bag of unrelated aerodynamic effects, when approaching supersonic speed. The term fell out of use in the 1950s, when aircraft started to routinely "break" the sound barrier.

Supersonic Objects

The tip of a bullwhip is thought to be the first man-made object to break the sound barrier, resulting in the telltale "crack" (actually a small sonic boom). The wave motion traveling through the bullwhip is what makes it capable of achieving supersonic speeds.

Most modern fighter aircraft are supersonic aircraft, but there have been supersonic passenger aircraft, namely Concorde and the Tupolev Tu-144. Both these passenger aircraft and some modern fighters are also capable of supercruise, a condition of sustained supersonic flight without the use of an afterburner. Due to its ability to supercruise for several hours and the relatively high frequency of flight over several decades, Concorde spent more time flying supersonically than all other aircraft combined by a considerable margin. Since Concorde's final retirement flight on November 26, 2003, there are no supersonic passenger aircraft left in service. Some large bombers, such as the Tupolev Tu-160 and Rockwell B-1 Lancer are also supersonic-capable.

Most modern firearm bullets are supersonic, with rifle projectiles often travelling at speeds approaching and in some cases well exceeding Mach 3.

Most spacecraft, most notably the Space Shuttle are supersonic at least during portions of their reentry, though the effects on the spacecraft are reduced by low air densities. During ascent, launch vehicles generally avoid going supersonic below 30 km (~98,400 feet) to reduce air drag.

Note that the speed of sound decreases somewhat with altitude, due to lower temperatures found there (typically up to 25 km). At even higher altitudes the temperature starts increasing, with the corresponding increase in the speed of sound.

When an inflated balloon is burst, the torn pieces of latex contract at supersonic speed, which contributes to the sharp and loud popping noise.

Supersonic Flight

Supersonic aerodynamics is simpler than subsonic aerodynamics because the airsheets at different points along the plane often can't affect each other. Supersonic jets and rocket vehicles require several times greater thrust to push through the extra aerodynamic drag experienced within the transonic region (around Mach 0.85–1.2). At these speeds aerospace engineers can gently guide air around the fuselage of the aircraft without producing new shock waves, but any change in cross area farther down the vehicle leads to shock waves along the body. Designers use the Supersonic area rule and the Whitcomb area rule to minimize sudden changes in size.

However, in practical applications, a supersonic aircraft must operate stably in both subsonic and supersonic profiles, hence aerodynamic design is more complex.

One problem with sustained supersonic flight is the generation of heat in flight. At high speeds aerodynamic heating can occur, so an aircraft must be designed to operate and function under very high temperatures. Duralumin, the traditional aircraft material, starts to lose strength and go into plastic deformation at relatively low temperatures, and is unsuitable for continuous use at speeds above Mach 2.2 to 2.4. Materials such as titanium and stainless steel allow operations at much higher temperatures. For example, the Lockheed SR-71 Blackbird jet could fly continuously at Mach 3.1 which could lead to temperatures on some parts of the aircraft getting above 315 °C (600 °F).

Another area of concern for sustained high-speed flight is engine operation. Jet engines create thrust by increasing the temperature of the air they ingest, and as the aircraft speeds up, friction and compression heat this air before it reaches the engines. The maximum allowable temperature of the exhaust is determined by the materials in the turbine at the rear of the engine, so as the aircraft speeds up, the difference in intake and exhaust temperature that the engine can create decreases, and the thrust along with it. Air cooling the turbine area to allow operations at higher temperatures was a key solution, one that continued to improve through the 1950s and on to this day.

Intake design was also a major issue. Normal jet engines can only ingest subsonic air, so for supersonic operation the air must be slowed down. Ramps or cones in the intake are used to create shock waves that slow the airflow before it reaches the engine. Doing so removes energy from the airflow, causing drag. The key to reducing this drag is to use multiple small oblique shock waves, but this was difficult because the angle they make inside the intake changes with Mach number. In order to efficiently operate across a range of speeds, the shock waves have to be "tuned."

An aircraft able to operate for extended periods at supersonic speeds has a potential range advantage over a similar design operating subsonically. Most of the drag an aircraft sees while speeding up to supersonic speeds occurs just below the speed of sound, due to an aerodynamic effect known as wave drag. An aircraft that can accelerate past this speed sees a significant drag decrease, and can fly supersonically with improved fuel economy. However, due to the way lift is generated supersonically, the lift-to-drag ratio of the aircraft as a whole drops, leading to lower range, offsetting or overturning this advantage.

The key to having low supersonic drag is to properly shape the overall aircraft to be long and thin, and close to a "perfect" shape, the von Karman ogive or Sears-Haack body. This has led to almost every supersonic cruising aircraft looking very similar to every other, with a very long and slender fuselage and large delta wings, cf. SR-71, Concorde, etc. Although not ideal for passenger aircraft, this shaping is quite adaptable for bomber use.

Hypersonic Speed

If travelling faster than sound (otherwise known as supersonic travel) wasn't speedy enough, we're now pushing aircrafts to go even faster: hypersonic speed.

The jump from supersonic to hypersonic travel isn't as clear cut as breaking the sound barrier – there's no sonic boom and instant shift in the aerodynamics – but is loosely defined as "Mach 5", or five times faster than the speed of sound, which would be 6,200km per hour at sea level.

Travel becomes hypersonic when temperatures get so hot that the air molecules become unstable and start losing electrons. At low temperatures, the air is made up of oxygen and nitrogen electrons, but at hypersonic speed, the air begins to change chemically and becomes an electrically-charged field.

Hypersonic speeds have been achieved before, most notably by the U.S. Air Force and NASA when test pilot William J. Knight set the record for fastest crewed flight in 1967 when he piloted the North American X-15 to Mach 6.72 (4,520 mph) at an altitude of 102,100 feet. In 2004, the uncrewed NASA X-43A set a speed record at Mach 9.6, or 7,310 mph at an altitude of about 110,000 feet.

Changes in the Motion of Objects at Hypersonic Speeds

To put it simply: everything we know about travel gets thrown out the window at hypersonic speeds.

"All of our understanding in aerodynamics falls over," says Dr Bruce. "Every equation that we use to describe sub sonic and supersonic flow and flight – so how we understand pressure force and temperature and the lift and drag of objects – doesn't work when the air starts changing at hypersonic speeds."

Theoretical understanding has dramatically improved in the last 50 years and we now have equations to describe aerodynamics at different Mach levels, but hypersonic travellers are now faced with an engineering challenge.

Characteristics of Flow

While the definition of hypersonic flow can be quite vague and is generally debatable (especially due to the absence of discontinuity between supersonic and hypersonic flows), a hypersonic flow may be characterized by certain physical phenomena that can no longer be analytically discounted as in supersonic flow. The peculiarity in hypersonic flows are as follows:

1. Shock layer
2. Aerodynamic heating
3. Entropy layer
4. Real gas effects
5. Low density effects
6. Independence of aerodynamic coefficients with Mach number

Small Shock stand-off Distance

As a body's Mach number increases, the density behind a bow shock generated by the body also increases, which corresponds to a decrease in volume behind the shock due to conservation of mass. Consequently, the distance between the bow shock and the body decreases at higher Mach numbers.

Entropy Layer

As Mach numbers increase, the entropy change across the shock also increases, which results in a strong entropy gradient and highly vortical flow that mixes with the boundary layer.

Viscous Interaction

A portion of the large kinetic energy associated with flow at high Mach numbers transforms into internal energy in the fluid due to viscous effects. The increase in internal energy is realized as an increase in temperature. Since the pressure gradient normal to the flow within a boundary layer is approximately zero for low to moderate hypersonic Mach numbers, the increase of temperature through the boundary layer coincides with a decrease in density. This causes the bottom of the boundary layer to expand, so that the boundary layer over the body grows thicker and can often merge with the shock wave near the body leading edge.

High-temperature Flow

High temperatures due to a manifestation of viscous dissipation cause non-equilibrium chemical flow properties such as vibrational excitation and dissociation and ionization of molecules resulting in convective and radiative heat-flux.

Classification of Mach Regimes

Although "subsonic" and "supersonic" usually refer to speeds below and above the local speed of sound respectively, aerodynamicists often use these terms to refer to particular ranges of Mach values. This occurs because a "transonic regime" exists around M=1 where approximations of the Navier–Stokes equations used for subsonic design no longer apply, partly because the flow locally exceeds M=1 even when the freestream Mach number is below this value.

The "supersonic regime" usually refers to the set of Mach numbers for which linearised theory may be used; for example, where the (air) flow is not chemically reacting and where heat transfer between air and vehicle may be reasonably neglected in calculations.

Generally, NASA defines "high" hypersonic as any Mach number from 10 to 25, and re-entry speeds as anything greater than Mach 25. Among the aircraft operating in this regime are the Space Shuttle and (theoretically) various developing spaceplanes.

Regimes

Hypersonic flow can be approximately separated into a number of regimes. The selection of these regimes is rough, due to the blurring of the boundaries where a particular effect can be found.

Perfect Gas

In this regime, the gas can be regarded as an ideal gas. Flow in this regime is still Mach number dependent. Simulations start to depend on the use of a constant-temperature wall, rather than the adiabatic wall typically used at lower speeds. The lower border of this region is around Mach 5, where ramjets become inefficient, and the upper border around Mach 10-12.

Two-temperature Ideal Gas

This is a subset of the perfect gas regime, where the gas can be considered chemically perfect, but the rotational and vibrational temperatures of the gas must be considered separately, leading to two temperature models. See particularly the modeling of supersonic nozzles, where vibrational freezing becomes important.

Dissociated Gas

In this regime, diatomic or polyatomic gases (the gases found in most atmospheres) begin to dissociate as they come into contact with the bow shock generated by the body. Surface catalysis plays a role in the calculation of surface heating, meaning that the type of surface material also has an effect on the flow. The lower border of this regime is where any component of a gas mixture first begins to dissociate in the stagnation point of a flow (which for nitrogen is around 2000 K). At the upper border of this regime, the effects of ionization start to have an effect on the flow.

Ionized Gas

In this regime the ionized electron population of the stagnated flow becomes significant, and the electrons must be modeled separately. Often the electron temperature is handled separately from the temperature of the remaining gas components. This region occurs for freestream flow velocities around 10–12 km/s. Gases in this region are modeled as non-radiating plasmas.

Radiation-dominated Regime

Above around 12 km/s, the heat transfer to a vehicle changes from being conductively dominated to radiatively dominated. The modeling of gases in this regime is split into two classes:

1. Optically thin: where the gas does not re-absorb radiation emitted from other parts of the gas,
2. Optically thick: where the radiation must be considered a separate source of energy.

The modeling of optically thick gases is extremely difficult, since, due to the calculation of the radiation at each point, the computation load theoretically expands exponentially as the number of points considered increases.

Wind Tunnel

Suppose you've just designed a gigantic new passenger airplane and now you want to test it out for real. You could spend millions of dollars building it out of shiny titanium metal and race it down the runway to see if it actually flies—but what if you got your calculations wrong? What if your airplane takes off for twenty seconds, then suddenly drops like a stone and lands on a city packed with 5 million people? It's not the best way for testing something so dangerous. That's why airplane designers try things out on the ground first using scale models in wind tunnels.

Wind tunnels are tube-shaped facilities that allow engineers to move air over a vehicle as if it were flying. They help researchers to learn more about how an aircraft will fly. NASA uses wind tunnels to test scale models of aircraft and spacecraft. Some wind tunnels are big enough to hold full-size versions of vehicles. By moving air around an object, the wind tunnel simulates the conditions of the object in flight.

Designing planes that will fly quickly, efficiently, and economically is all about making air flow smoothly over their wings and past their tube-like bodies. This is called the science of aerodynamics. Once a plane's up in the air, there's no easy way to see how air is moving past it (though an experienced test pilot will have a good idea what might be causing problems). If there's a major design defect, the plane won't get into the air at all. That's why every modern spacecraft and airplane is tested on the ground first in a wind tunnel: a pipe-like building through which air blasts at very high speed.

The basic idea of a wind tunnel is simple: if you can't move the plane through the air, why not move the air past the plane instead? From a scientific point of view, it's exactly the same. If a plane drags (causes air resistance) when it soars through the sky, air will drag in exactly the same way when you fire it past a stationary model of the plane on the ground.

There's nothing to stop you building a super-giant wind tunnel and testing a life-sized model of your plane—and, indeed, the American space agency NASA does have wind tunnels like this. But most of the time it's much cheaper to use a small, scale model of the plane in a much smaller wind tunnel.

Measurement of Aerodynamic Forces

Air velocity and pressures are measured in several ways in wind tunnels.

Air velocity through the test section is determined by Bernoulli's principle. Measurement of the dynamic pressure, the static pressure, and (for compressible flow only) the temperature rise in the airflow. The direction of airflow around a model can be determined by tufts of yarn attached to the aerodynamic surfaces. The direction of airflow approaching a surface can be visualized by mounting threads in the airflow ahead of and aft of the test model. Smoke or bubbles of liquid can be introduced into the airflow upstream of the test model, and their path around the model can be photographed.

Aerodynamic forces on the test model are usually measured with beam balances, connected to the test model with beams, strings, or cables.

The pressure distributions across the test model have historically been measured by drilling many small holes along the airflow path, and using multi-tube manometers to measure the pressure at each hole. Pressure distributions can more conveniently be measured by the use of pressure-sensitive paint, in which higher local pressure is indicated by lowered fluorescence of the paint at that point. Pressure distributions can also be conveniently measured by the use of pressure-sensitive pressure belts, a recent development in which multiple ultra-miniaturized pressure sensor modules are integrated into a flexible strip. The strip is attached to the aerodynamic surface with tape, and it sends signals depicting the pressure distribution along its surface.

Pressure distributions on a test model can also be determined by performing a wake survey, in which either a single pitot tube is used to obtain multiple readings downstream of the test model, or a multiple-tube manometer is mounted downstream and all its readings are taken.

The aerodynamic properties of an object can not all remain the same for a scaled model. However, by observing certain similarity rules, a very satisfactory correspondence between the aerodynamic properties of a scaled model and a full-size object can be achieved. The choice of similarity parameters depends on the purpose of the test, but the most important conditions to satisfy are usually:

- Geometric similarity: all dimensions of the object must be proportionally scaled;
- Mach number: the ratio of the airspeed to the speed of sound should be identical for the scaled model and the actual object (having identical Mach number in a wind tunnel and around the actual object is -not- equal to having identical airspeeds);
- Reynolds number: the ratio of inertial forces to viscous forces should be kept. This parameter is difficult to satisfy with a scaled model and has led to development of pressurized and cryogenic wind tunnels in which the viscosity of the working fluid can be greatly changed to compensate for the reduced scale of the model.

In certain particular test cases, other similarity parameters must be satisfied, such as e.g. Froude number.

How it Works

Air is blown or sucked through a duct equipped with a viewing port and instrumentation where models or geometrical shapes are mounted for study. Typically the air is moved through the tunnel using a series of fans. For very large wind tunnels several meters in diameter, a single large fan is not practical, and so instead an array of multiple fans are used in parallel to provide sufficient airflow. Due to the sheer volume and speed of air movement required, the fans may be powered by stationary turbofan engines rather than electric motors.

Six-element external balance below the Kirsten Wind Tunnel.

The airflow created by the fans that is entering the tunnel is itself highly turbulent due to the fan blade motion (when the fan is blowing air into the test section – when it is sucking air out of the test section downstream, the fan-blade turbulence is not a factor), and so is not directly useful for accurate measurements. The air moving through the tunnel needs to be relatively turbulence-free and laminar. To correct this problem, closely spaced vertical and horizontal air vanes are used to smooth out the turbulent airflow before reaching the subject of the testing.

Due to the effects of viscosity, the cross-section of a wind tunnel is typically circular rather than square, because there will be greater flow constriction in the corners of a square tunnel that can make the flow turbulent. A circular tunnel provides a smoother flow.

The inside facing of the tunnel is typically as smooth as possible, to reduce surface drag and turbulence that could impact the accuracy of the testing. Even smooth walls induce some drag into the airflow, and so the object being tested is usually kept near the center of the tunnel, with an empty buffer zone between the object and the tunnel walls. There are correction factors to relate wind tunnel test results to open-air results.

The lighting is usually embedded into the circular walls of the tunnel and shines in through windows. If the light were mounted on the inside surface of the tunnel in a conventional manner, the light bulb would generate turbulence as the air blows around it. Similarly, observation is usually done through transparent portholes into the tunnel. Rather than simply being flat discs, these lighting and observation windows may be curved to match the cross-section of the tunnel and further reduce turbulence around the window.

Various techniques are used to study the actual airflow around the geometry and compare it with theoretical results, which must also take into account the Reynolds number and Mach number for the regime of operation.

Pressure Measurements

Pressure across the surfaces of the model can be measured if the model includes pressure taps. This can be useful for pressure-dominated phenomena, but this only accounts for normal forces on the body.

Force and Moment Measurements

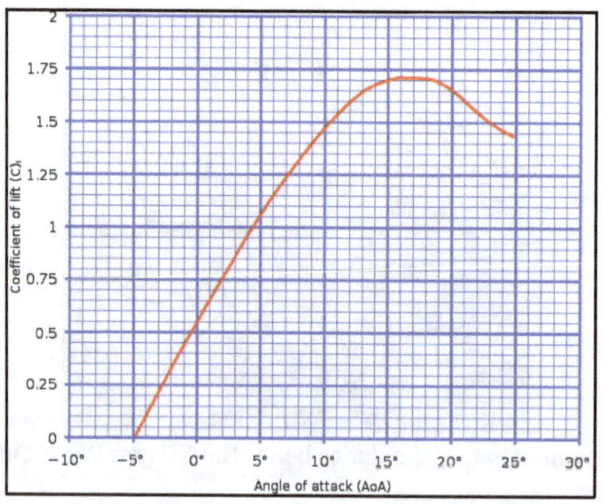

A typical lift coefficient versus angle of attack curve.

With the model mounted on a force balance, one can measure lift, drag, lateral forces, yaw, roll, and pitching moments over a range of angle of attack. This allows one to produce common curves such as lift coefficient versus angle of attack.

Note that the force balance itself creates drag and potential turbulence that will affect the model and introduce errors into the measurements. The supporting structures are therefore typically smoothly shaped to minimize turbulence.

Flow Visualization

Because air is transparent it is difficult to directly observe the air movement itself. Instead, multiple methods of both quantitative and qualitative flow visualization methods have been developed for testing in a wind tunnel.

Qualitative Methods

- Smoke.
- Tufts are applied to a model and remain attached during testing. Tufts can be used to gauge air flow patterns and flow separation.

- Evaporating suspensions are simply a mixture of some sort or fine powder, talc, or clay mixed into a liquid with a low latent heat of evaporation. When the wind is turned on the liquid quickly evaporates, leaving behind the clay in a pattern characteristic of the air flow.
- Oil: When oil is applied to the model surface it can clearly show the transition from laminar to turbulent flow as well as flow separation.
- Fog (usually from water particles) is created with an ultrasonic piezoelectric nebulizer. The fog is transported inside the wind tunnel (preferably of the closed circuit and closed test section type). An electrically heated grid is inserted before the test section, which evaporates the water particles at its vicinity, thus forming fog sheets. The fog sheets function as streamlines over the test model when illuminated by a light sheet.
- Sublimation: If the air movement in the tunnel is sufficiently non-turbulent, a particle stream released into the airflow will not break up as the air moves along, but stay together as a sharp thin line. Multiple particle streams released from a grid of many nozzles can provide a dynamic three-dimensional shape of the airflow around a body. As with the force balance, these injection pipes and nozzles need to be shaped in a manner that minimizes the introduction of turbulent airflow into the airstream.

High-speed turbulence and vortices can be difficult to see directly, but strobe lights and film cameras or high-speed digital cameras can help to capture events that are a blur to the naked eye.

High-speed cameras are also required when the subject of the test is itself moving at high speed, such as an airplane propeller. The camera can capture stop-motion images of how the blade cuts through the particulate streams and how vortices are generated along the trailing edges of the moving blade.

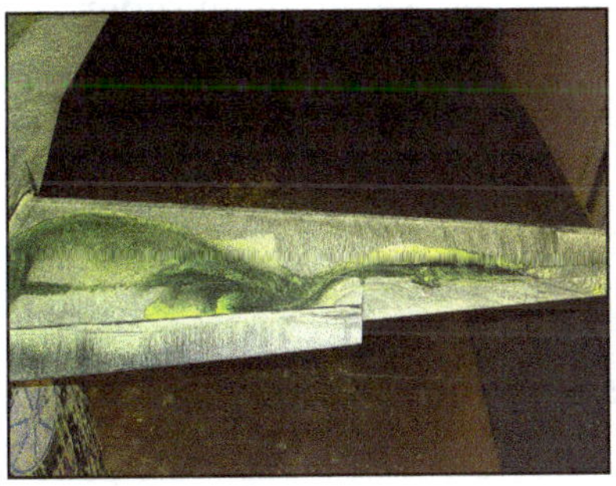

China clay on a wing in the Kirsten Wind Tunnel showing reverse and span-wise flow.

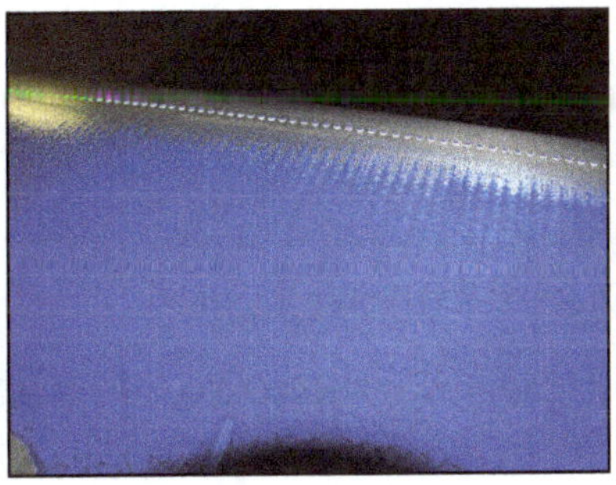

Oil flow visible on a straight wing in the Kirsten Wind Tunnel. Trip dots can be seen near the leading edge.

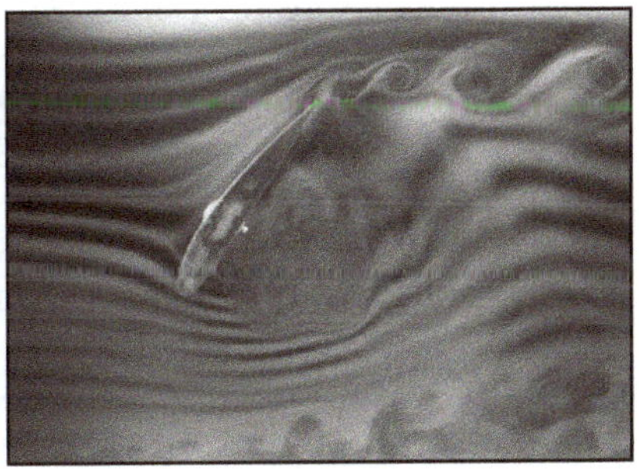

Fog (water particle) wind tunnel visualization of a NACA 4412 airfoil at a low-speed flow (Re=20.000).

Classification

There are many different kinds of wind tunnels. They are typically classified by the range of speeds that are achieved in the test section, as follows:

- Low-speed wind tunnel
- High speed wind tunnel

- Subsonic and transonic wind tunnel
- Supersonic wind tunnel
- Hypersonic wind tunnel

Wind tunnels are also classified by the orientation of air flow in the test section with respect to gravity. Typically they are oriented horizontally, as happens during level flight. A different class of wind tunnels are oriented vertically so that gravity can be balanced by drag instead of lift, and these have become a popular form of recreation for simulating sky-diving:

- Vertical wind tunnel

Wind tunnels are also classified based on their main use. For those used with land vehicles such as cars and trucks the type of floor aerodynamics is also important. These vary from stationary floors through to full moving floors, with smaller moving floors and some attempt at boundary level control also being important.

Aeronautical Wind Tunnels

The main subcategories in the aeronautical wind tunnels are:

- High Reynolds Number Tunnels

Reynolds number is one of the governing similarity parameters for the simulation of flow in a wind tunnel. For mach number less than 0.3, it is the primary parameter that governs the flow characteristics. There are three main ways to simulate high Reynolds number, since it is not practical to obtain full scale Reynolds number by use of a full scale vehicle.

- Pressurised tunnels: Here test gases are pressurised to increase the Reynolds number.
- Heavy gas tunnels: Heavier gases like freon and R-134a are used as test gases. The transonic dynamics tunnel at NASA Langley is an example of such a tunnel.
- Cryogenic tunnels: Here test gas is cooled down to increase the Reynolds number. The European transonic wind tunnel uses this technique.
- High-altitude tunnels: These are designed to test the effects of shock waves against various aircraft shapes in near vacuum. In 1952 the University of California constructed the first two high-altitude wind tunnels: one for testing objects at 50 to 70 miles above the earth and the second for tests at 80 to 200 miles above the earth.
- V/STOL Tunnels.

V/STOL tunnels require large cross section area, but only small velocities. Since power varies with the cube of velocity, the power required for the operation is also less. An example of a V/STOL tunnel is the NASA Langley 14' x 22' tunnel.

- Spin Tunnels.

Aircraft have a tendency to go to spin when they stall. These tunnels are used to study that phenomenon.

Automotive Tunnels

Automotive wind tunnels fall into two categories:

- External flow tunnels are used to study the external flow through the chassis
- Climatic tunnels are used to evaluate the performance of door systems, braking systems, etc. under various climatic conditions. Most of the leading automobile manufacturers have their own climatic wind tunnels

Wunibald Kamm built the first full-scale wind tunnel for motor vehicles.

For external flow tunnels various systems are used to compensate for the effect of the boundary layer on the road surface, including systems of moving belts under each wheel and the body of the car (5 or 7 belt systems) or one large belt under the entire car, or other methods of boundary layer control such as scoops or perforations to suck it away.

Aeroacoustic Tunnels

These tunnels are used in the studies of noise generated by flow and its suppression.

Aquadynamic Flume

Vertical wind tunnel T-105 at Central Aerohydrodynamic Institute, Moscow, built in 1941 for aircraft testing.

The aerodynamic principles of the wind tunnel work equally on watercraft, except the water is more viscous and so sets greater forces on the object being tested. A looping flume is typically used for underwater aquadynamic testing. The interaction between two different types of fluids means that pure wind tunnel testing is only partly relevant. However, a similar sort of research is done in a towing tank.

Low-speed Oversize Liquid Testing

Air is not always the best test medium for studying small-scale aerodynamic principles, due to the speed of the air flow and airfoil movement. A study of fruit fly wings designed to understand how the wings produce lift was performed using a large tank of mineral oil and wings 100 times larger than actual size, in order to slow down the wing beats and make the vortices generated by the insect wings easier to see and understand.

Fan Testing

Wind tunnel tests are also performed to precisely measure the air movement of fans at a specific pressure. By determining the environmental circumstances during measurement, and by revising the air-tightness afterwards, the standardization of the data is ensured.

There are two possible ways of measurement: a complete fan, or an impeller on a hydraulic installation. Two measuring tubes enable measurements of lower air currents (< 30.000 m^3/h) as well as higher air currents (< 60.000 m^3/h). The determination of the Q/h curve of the fan is one of the main objectives. To determine this curve (and to define other parameters) air technical, mechanical as well as electro technical data are measured.

Air technical:

- Static pressure difference (Pa)
- Amount of moved air (m^3/h)
- Average air speed (m/s)
- Specific efficiency (W/1000m^3/h)
- Efficiency

Electro technical:

- Tension (V)
- Current (A)
- Cos φ
- Admitted power (W) fan / impeller
- Rotations per minute (RPM)

The measurement can take place on the fan or in the application in which the fan is used.

Wind Engineering Testing

In wind engineering, wind tunnel tests are used to measure the velocity around, and forces or pressures upon structures. Very tall buildings, buildings with unusual or complicated shapes (such as a tall building with a parabolic or a hyperbolic shape), cable suspension bridges or cable stayed bridges are analyzed in specialized atmospheric boundary layer wind tunnels. These feature a long upwind section to accurately represent the wind speed and turbulence profile acting on the structure. Wind tunnel tests provide the necessary design pressure measurements in use of the dynamic analysis and control of tall buildings.

Importance of Wind Tunnels in Spacecraft

NASA also uses wind tunnels to test spacecraft and rockets. These vehicles are made to operate in space. Space has no atmosphere. Spacecraft and rockets have to travel through the atmosphere to get to space. Vehicles that take humans into space also must come back through the atmosphere to Earth.

Wind tunnels are important. NASA uses them to test the Orion spacecraft and the Space Launch System rockets. These rockets are called the SLS. Orion and SLS are new vehicles. They will take astronauts into space. NASA must test the systems in wind tunnels to see if they are safe to fly. And NASA must see what happens when Orion comes back to Earth through the atmosphere.

Wind tunnels can even help engineers design spacecraft to work on other worlds. Mars has a thin atmosphere. It is important to know what the Martian atmosphere will do to vehicles that are landing there. Spacecraft designs and parachutes are tested in wind tunnels set up to be like the Martian atmosphere.

Types of Wind Tunnels used by NASA

NASA has many different types of wind tunnels. They are located at NASA centers all around the country. The wind tunnels come in a lot of sizes. Some are only a few inches square, and some are large enough to test a full-size airplane. Some wind tunnels test aircraft at very slow speeds. But some wind tunnels are made to test at hypersonic speeds. That is more than 4,000 miles per hour.

References

- Williamson, C. H. K.; Govardhan, R (2004), "Vortex-induced vibrations", Annual Review of Fluid Mechanics, 36: 413–455, Bibcode:2004AnRFM..36..413W, doi:10.1146/annurev.fluid.36.050802.122128
- G. Falkovich (2011). Fluid Mechanics (A short course for physicists). Cambridge University Press. ISBN 978-1-107-00575-4
- Spalart, P. R. (2000), Strategies for turbulence modeling and simulations, 21 (3), International Journal of Heat and Fluid Flow, p. 252
- Hernandez-Gomez, J J; Marquina, V; Gomez, R W (25 July 2013). "On the performance of Usain Bolt in the 100 m sprint". Eur. J. Phys. IOP. 34 (5): 1227. Bibcode:2013EJPh...34.1227H. doi:10.1088/0143-0807/34/5/1227. Retrieved 23 April 2020
- Marchaj, C. A. (2003). Sail performance : techniques to maximise sail power (Rev. ed.). London: Adlard Coles Nautical. pp. 147 figure 127 lift vs drag polar curves. ISBN 978-0-7136-6407-2
- "U.S Centennial of Flight Commission – Sir George Cayley". Archived from the original on 20 September 2008. Retrieved 22, July 2020
- Smith, N. F., "Bernoulli and Newton in Fluid Mechanics", The Physics Teacher, 10 (8): 451, Bibcode:1972PhTea..10..451S, doi:10.1119/1.2352317

- Smith, Hubert (1992). The Illustrated Guide to Aerodynamics (2nd ed.). New York: TAB Books. pp. 24–27. ISBN 0-8306-3901-2
- Kármán, Th V. "Über laminare und turbulente Reibung." ZAMM-Journal of Applied Mathematics and Mechanics/Zeitschrift für Angewandte Mathematik und Mechanik 1.4 : 233-252
- Jr., John D. Anderson,. "Research in Supersonic Flight and the Breaking of the Sound Barrier". history.nasa.gov. Retrieved 14 April 2020
- Young, A.D. (1989). Boundary layers (1st publ. ed.). Washington, DC: American Institute of Aeronautics and Astronautics. ISBN 0930403576
- Wille, R.; Fernholz, H., "Report on the first European Mechanics Colloquium, on the Coanda effect", J. Fluid Mech., 23 (4): 801–819, Bibcode:1965JFM....23..801W, doi:10.1017/s0022112065001702

5

Pressure Measurement for Aircraft

There are numerous devices and methods which are used for the measurement of pressure in aerospace engineering. A few of them are McLeod gauge, Pirani gauge, thermocouple gauge and hot-filament ionization gauge. The topics elaborated in this chapter will help in gaining a better perspective about these devices used for pressure measurement.

Pressure Measurement

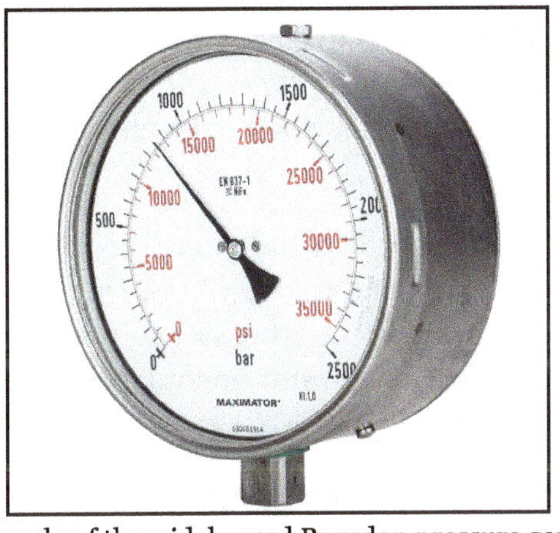

Example of the widely used Bourdon pressure gauge.

Many techniques have been developed for the measurement of pressure and vacuum. Instruments used to measure and display pressure in an integral unit are called pressure gauges or vacuum gauges. A manometer is a good example as it uses a column of liquid to both measure and indicate pressure. Likewise the widely used Bourdon gauge is a mechanical device which both measures and indicates, and is probably the best known type of gauge.

A vacuum gauge is an absolute pressure gauge used to measure the pressures lower than the ambient atmospheric pressure.

Other methods of pressure measurement involve sensors which can transmit the pressure reading to a remote indicator or control system (telemetry).

Absolute, Gauge and Differential Pressures — Zero Reference

Everyday pressure measurements, such as for vehicle tire pressure, are usually made relative to ambient air pressure. In other cases measurements are made relative to a vacuum or to some other specific reference. When distinguishing between these zero references, the following terms are used:

- Absolute pressure is zero-referenced against a perfect vacuum, using an absolute scale, so it is equal to gauge pressure plus atmospheric pressure.

- Gauge pressure is zero-referenced against ambient air pressure, so it is equal to absolute pressure minus atmospheric pressure. Negative signs are usually omitted. To distinguish a negative pressure, the value may be appended with the word "vacuum" or the gauge may be labeled a "vacuum gauge." These are further divided into two subcategories: high and low vacuum (and sometimes ultra-high vacuum). The applicable pressure ranges of many of the techniques used to measure vacuums have an overlap. Hence, by combining several different types of gauge, it is possible to measure system pressure continuously from 10 mbar down to 10^{-11} mbar.

- Differential pressure is the difference in pressure between two points.

The zero reference in use is usually implied by context, and these words are added only when clarification is needed. Tire pressure and blood pressure are gauge pressures by convention, while atmospheric pressures, deep vacuum pressures, and altimeter pressures must be absolute.

For most working fluids where a fluid exists in a closed system, gauge pressure measurement prevails. Pressure instruments connected to the system will indicate pressures relative to the current atmospheric pressure. The situation changes when extreme vacuum pressures are measured; absolute pressures are typically used instead.

Differential pressures are commonly used in industrial process systems. Differential pressure gauges have two inlet ports, each connected to one of the volumes whose pressure is to be monitored. In effect, such a gauge performs the mathematical operation of subtraction through mechanical means, obviating the need for an operator or control system to watch two separate gauges and determine the difference in readings.

Moderate vacuum pressure readings can be ambiguous without the proper context, as they may represent absolute pressure or gauge pressure without a negative sign. Thus a vacuum of 26 inHg gauge is equivalent to an absolute pressure of 30 inHg (typical atmospheric pressure) – 26 inHg = 4 inHg.

Atmospheric pressure is typically about 100 kPa at sea level, but is variable with altitude and weather. If the absolute pressure of a fluid stays constant, the gauge pressure of the same fluid will vary as atmospheric pressure changes. For example, when a car drives up a mountain, the (gauge) tire pressure goes up because atmospheric pressure goes down. The absolute pressure in the tire is essentially unchanged.

Using atmospheric pressure as reference is usually signified by a *g* for gauge after the pressure unit, e.g. 70 psig, which means that the pressure measured is the total pressure minus atmospheric pressure. There are two types of gauge reference pressure: vented gauge (vg) and sealed gauge (sg).

A vented gauge pressure transmitter for example allows the outside air pressure to be exposed to the negative side of the pressure sensing diaphragm, via a vented cable or a hole on the side of the device, so that it always measures the pressure referred to ambient barometric pressure. Thus a vented gauge reference pressure sensor should always read zero pressure when the process pressure connection is held open to the air.

A sealed gauge reference is very similar except that atmospheric pressure is sealed on the negative side of the diaphragm. This is usually adopted on high pressure ranges such as hydraulics where atmospheric pressure changes will have a negligible effect on the accuracy of the reading, so venting is not necessary. This also allows some manufacturers to provide secondary pressure containment as an extra precaution for pressure equipment safety if the burst pressure of the primary pressure sensing diaphragm is exceeded.

There is another way of creating a sealed gauge reference and this is to seal a high vacuum on the reverse side of the sensing diaphragm. Then the output signal is offset so the pressure sensor reads close to zero when measuring atmospheric pressure.

A sealed gauge reference pressure transducer will never read exactly zero because atmospheric pressure is always changing and the reference in this case is fixed at 1 bar.

To produce an absolute pressure sensor the manufacturer will seal a high vacuum behind the sensing diaphragm. If the process pressure connection of an absolute pressure transmitter is open to the air, it will read the actual barometric pressure.

Units

Pressure units						
v · t · e	**Pascal**	**Bar**	**Technical atmosphere**	**Standard atmosphere**	**Torr**	**Pounds per square inch**
	(Pa)	**(bar)**	**(at)**	**(atm)**	**(Torr)**	**(psi)**
1 Pa	≡ 1 N/m²	10^{-5}	1.0197×10^{-5}	9.8692×10^{-6}	7.5006×10^{-3}	1.450377×10^{-4}
1 bar	10^5	≡ 100 kPa ≡ 10^6 dyn/cm²	1.0197	0.98692	750.06	14.50377
1 at	9.80665×10^4	0.980665	≡ 1 kp/cm²	0.9678411	735.5592	14.22334
1 atm	1.01325×10^5	1.01325	1.0332	1	≡ 760	14.69595
1 Torr	133.3224	1.333224×10^{-3}	1.359551×10^{-3}	≡ 1/760 ≈ 1.315789×10^{-3}	≡ 1 Torr ≈ 1 mmHg	1.933678×10^{-2}
1 psi	6.8948×10^3	6.8948×10^{-2}	7.03069×10^{-2}	6.8046×10^{-2}	51.71493	≡ 1 lbf /in²

The SI unit for pressure is the pascal (Pa), equal to one newton per square metre ($N \cdot m^{-2}$ or $kg \cdot m^{-1} \cdot s^{-2}$). This special name for the unit was added in 1971; before that, pressure in SI was expressed in units such as $N \cdot m^{-2}$. When indicated, the zero reference is stated in parenthesis following the unit, for example 101 kPa (abs). The pound per square inch (psi) is still in widespread use in the US and Canada, for measuring, for instance, tire pressure. A letter is often appended to the psi unit to indicate the measurement's zero reference; psia for absolute, psig for gauge, psid for differential, although this practice is discouraged by the NIST.

Because pressure was once commonly measured by its ability to displace a column of liquid in a manometer, pressures are often expressed as a depth of a particular fluid (*e.g.*, inches of water). Manometric measurement is the subject of pressure head calculations. The most common choices for a manometer's fluid are mercury (Hg) and water; water is nontoxic and readily available, while mercury's density allows for a shorter column (and so a smaller manometer) to measure a given pressure. The abbreviation "W.C." or the words "water column" are often printed on gauges and measurements that use water for the manometer.

Fluid density and local gravity can vary from one reading to another depending on local factors, so the height of a fluid column does not define pressure precisely. So measurements in "millimetres of mercury" or "inches of mercury" can be converted to SI units as long as attention is paid to the local factors of fluid density and gravity. Temperature fluctuations change the value of fluid density, while location can affect gravity.

Although no longer preferred, these manometric units are still encountered in many fields. Blood pressure is measured in millimetres of mercury in most of the world, and lung pressures in centimeters of water are still common, as in settings for CPAP machines. Natural gas pipeline pressures are measured in inches of water, expressed as "inches W.C." Scuba divers often use a manometric rule of thumb: the pressure exerted by ten meters depth of sea water ("10 msw") is approximately equal to one atmosphere. In vacuum systems, the units torr (millimeter of mercury), micron (micrometer of mercury), and inch of mercury (inHg) are most commonly used. Torr and micron usually indicates an absolute pressure, while inHg usually indicates a gauge pressure.

Atmospheric pressures are usually stated using hectopascal (hPa), kilopascal (kPa), millibar (mbar) or atmospheres (atm). In American and Canadian engineering, stress is often measured in kip. Note that stress is not a true pressure since it is not scalar. In the cgs system the unit of pressure was the barye (ba), equal to 1 dyn·cm^{-2}. In the mts system, the unit of pressure was the pieze, equal to 1 sthene per square metre.

Many other hybrid units are used such as mmHg/cm^2 or grams-force/cm^2 (sometimes as [[kg/cm^2]] without properly identifying the force units). Using the names kilogram, gram, kilogram-force, or gram-force (or their symbols) as a unit of force is prohibited in SI; the unit of force in SI is the newton (N).

Static and Dynamic Pressure

Static pressure is uniform in all directions, so pressure measurements are independent of direction in an immovable (static) fluid. Flow, however, applies additional pressure on surfaces perpendicular to the flow direction, while having little impact on surfaces parallel to the flow direction. This directional component of pressure in a moving (dynamic) fluid is called dynamic pressure. An instrument facing the flow direction measures the sum of the static and dynamic pressures; this measurement is called the total pressure or stagnation pressure. Since dynamic pressure is referenced to static pressure, it is neither gauge nor absolute; it is a differential pressure.

While static gauge pressure is of primary importance to determining net loads on pipe walls, dynamic pressure is used to measure flow rates and airspeed. Dynamic pressure can be measured by taking the differential pressure between instruments parallel and perpendicular to the flow. Pitot-static tubes, for example perform this measurement on airplanes to determine airspeed. The presence of the measuring instrument inevitably acts to divert flow and create turbulence, so its shape is critical to accuracy and the calibration curves are often non-linear.

Applications

- Altimeter
- Barometer
- MAP sensor
- Pitot tube
- Sphygmomanometer

Instruments

Many instruments have been invented to measure pressure, with different advantages and disadvantages. Pressure range, sensitivity, dynamic response and cost all vary by several orders of magnitude from one instrument design to the next. The oldest type is the liquid column (a vertical tube filled with mercury) manometer invented by Evangelista Torricelli in 1643. The U-Tube was invented by Christiaan Huygens in 1661.

Hydrostatic

Hydrostatic gauges (such as the mercury column manometer) compare pressure to the hydrostatic force per unit area at the base of a column of fluid. Hydrostatic gauge measurements are independent of the type of gas being measured, and can be designed to have a very linear calibration. They have poor dynamic response.

Piston

Piston-type gauges counterbalance the pressure of a fluid with a spring (for example tire-pressure gauges of comparatively low accuracy) or a solid weight, in which case it is known as a deadweight tester and may be used for calibration of other gauges.

Liquid Column (Manometer)

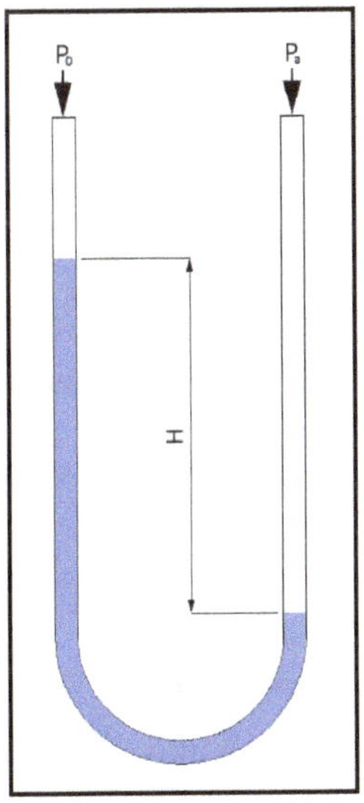

The difference in fluid height in a liquid column manometer is proportional to the pressure difference.

By using the pressure head equation, liquids can be used for instrumentation where gravity is present. Liquid column gauges consist of a vertical column of liquid in a tube that has ends which are exposed to different pressures. The column will rise or fall until its weight (a force applied due to gravity) is in equilibrium with the pressure differential between the two ends of the tube (a force applied due to fluid pressure). A very simple version is a U-shaped tube half-full of liquid, one side of which is connected to the region of interest while the reference pressure (which might be the atmospheric pressure or a vacuum) is applied to the other. The difference in liquid level represents the applied pressure. The pressure exerted by a column of fluid of height h and density ρ is given by the hydrostatic pressure equation, $P = hg\rho$. Therefore, the pressure difference between the applied pressure P_a and the reference pressure P_0 in a U-tube manometer can be found by solving $P_a - P_0 = hg\rho$. In other words, the pressure on either end of the liquid (shown in blue in the figure) must be balanced (since the liquid is static) and so $P_a = P_0 + hg\rho$.

In most liquid column measurements, the result of the measurement is the height, h, expressed typically in mm, cm, or inches. The h is also known as the pressure head. When expressed as a pressure head, pressure is specified in units of length and the measurement fluid must be specified. When accuracy is critical, the temperature of the measurement fluid must likewise be specified, because liquid density is a function of temperature. So, for example, pressure head might be written "742.2 mm_{Hg}" or "4.2 in_{H2O} at 59 °F" for measurements taken with mercury or water as the manometric fluid, respectively. The word "gauge" or "vacuum" may be added to such a measurement to distinguish between a pressure above or below the atmospheric pressure. Both mm of mercury and inches of water are common pressure heads which can be converted to S.I. units of pressure using unit conversion and the above formulas.

If the fluid being measured is significantly dense, hydrostatic corrections may have to be made for the height between the moving surface of the manometer working fluid and the location where the pressure measurement is desired except when measuring differential pressure of a fluid (for example across an orifice plate or venturi), in which case the density ρ should be corrected by subtracting the density of the fluid being measured.

Although any fluid can be used, mercury is preferred for its high density (13.534 g/cm³) and low vapour pressure. For low pressure differences, light oil or water are commonly used (the latter giving rise to units of measurement such as inches water gauge and millimetres H_2O. Liquid-column pressure gauges have a highly linear calibration. They have poor dynamic response because the fluid in the column may react slowly to a pressure change.

When measuring vacuum, the working liquid may evaporate and contaminate the vacuum if its vapor pressure is too high. When measuring liquid pressure, a loop filled with gas or a light fluid can isolate the liquids to prevent them from mixing but this can be unnecessary, for example when mercury is used as the manometer fluid to measure differential pressure of a fluid such as water. Simple hydrostatic gauges can measure pressures ranging from a few Torr (a few 100 Pa) to a few atmospheres. (Approximately 1,000,000 Pa)

A single-limb liquid-column manometer has a larger reservoir instead of one side of the U-tube and has a scale beside the narrower column. The column may be inclined to further amplify the liquid movement. Based on the use and structure following type of manometers are used:

1. Simple Manometer
2. Micromanometer
3. Differential manometer
4. Inverted differential manometer

McLeod Gauge

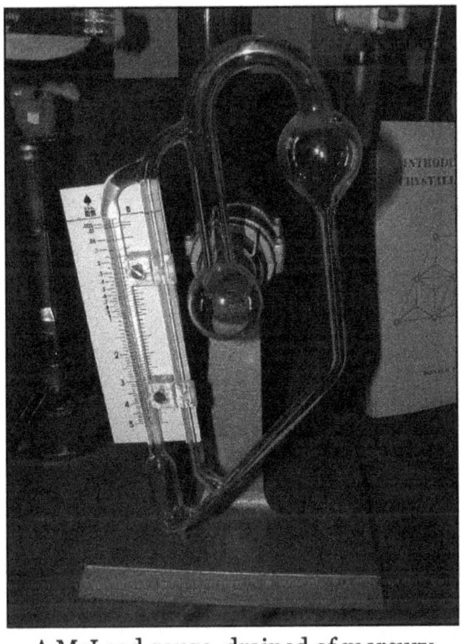

A McLeod gauge, drained of mercury.

A McLeod gauge isolates a sample of gas and compresses it in a modified mercury manometer until the pressure is a few millimetres of mercury. The technique is slow and unsuited to continual monitoring, but is capable of good accuracy. Unlike other manometer gauges, the McLeod gauge reading is dependent on the composition of the gas since the interpretation relies on the sample compressing as an ideal gas. Due to the compression process, the McLeod gauge completely ignores partial pressures from non-ideal vapors that condense, such as pump oils, mercury, and even water if compressed enough.

Useful range: from around 10^{-4} torr (roughly 10^{-2} Pa) to vacuums as high as 10^{-6} Torr (0.1 mPa),

0.1 mPa is the lowest direct measurement of pressure that is possible with current technology. Other vacuum gauges can measure lower pressures, but only indirectly by measurement of other pressure-controlled properties. These indirect measurements must be calibrated to SI units via a direct measurement, most commonly a McLeod gauge.

Aneroid

Aneroid gauges are based on a metallic pressure-sensing element that flexes elastically under the effect of a pressure difference across the element. "Aneroid" means "without fluid," and the term originally distinguished these gauges from the hydrostatic gauges described above. However, aneroid gauges can be used to measure the pressure of a liquid as well as a gas, and they are not the only type of gauge that can operate without fluid. For this reason, they are often called mechanical gauges in modern language. Aneroid gauges are not dependent on the type of gas being measured, unlike thermal and ionization gauges, and are less likely to contaminate the system than hydrostatic gauges. The pressure sensing element may be a Bourdon tube, a diaphragm, a capsule, or a set of bellows, which will change shape in response to the pressure of the region in question. The deflection of the pressure sensing element may be read by a linkage connected to a needle, or it may be read by a secondary transducer. The most common secondary transducers in modern vacuum gauges measure a change in capacitance due to the mechanical deflection. Gauges that rely on a change in capacitance are often referred to as capacitance manometers.

Bourdon

Membrane-type manometer.

The Bourdon pressure gauge uses the principle that a flattened tube tends to straighten or regain its circular form in cross-section when pressurized. Although this change in cross-section may be hardly noticeable, and thus involving moderate stresses within the elastic range of easily workable materials, the strain of the material of the tube is magnified by forming the tube into a C shape or even a helix, such that the entire tube tends to straighten out or uncoil, elastically, as it is pressurized. Eugene Bourdon patented his gauge in France in 1849, and it was widely adopted because of its superior sensitivity, linearity, and accuracy; Edward Ashcroft purchased Bourdon's American patent rights in 1852 and became a major manufacturer of gauges. Also in 1849, Bernard Schaeffer in Magdeburg, Germany patented a successful diaphragm pressure gauge, which, together with the Bourdon gauge, revolutionized pressure measurement in industry. But in 1875 after Bourdon's patents expired, his company Schaeffer and Budenberg also manufactured Bourdon tube gauges.

An original 19th century Eugene Bourdon compound gauge, reading pressure both below and above ambient with great sensitivity.

In practice, a flattened thin-wall, closed-end tube is connected at the hollow end to a fixed pipe containing the fluid pressure to be measured. As the pressure increases, the closed end moves in an arc, and this motion is converted into the rotation of a (segment of a) gear by a connecting link that is usually adjustable. A small-diameter pinion gear is on the pointer shaft, so the motion is magnified further by the gear ratio. The positioning of the indicator card behind the pointer, the initial pointer shaft position, the linkage length and initial position, all provide means to calibrate the pointer to indicate the desired range of pressure for variations in the behavior of the Bourdon tube itself. Differential pressure can be measured by gauges containing two different Bourdon tubes, with connecting linkages.

Bourdon tubes measure gauge pressure, relative to ambient atmospheric pressure, as opposed to absolute pressure; vacuum is sensed as a reverse motion. Some aneroid barometers use Bourdon tubes closed at both ends. When the measured pressure is rapidly pulsing, such as when the gauge is near a reciprocating pump, an orifice restriction in the connecting pipe is frequently used to avoid unnecessary wear on the gears and provide an average reading; when the whole gauge is subject to mechanical vibration, the entire case including the pointer and indicator card can be filled with an oil or glycerin. Tapping on the face of the gauge is not recommended as it will tend to falsify actual readings initially presented by the gauge. The Bourdon tube is separate from the face of the gauge and thus has no effect on the actual reading of pressure. Typical high-quality modern gauges provide an accuracy of ±2% of span, and a special high-precision gauge can be as accurate as 0.1% of full scale.

In the following illustrations the transparent cover face of the pictured combination pressure and vacuum gauge has been removed and the mechanism removed from the case. This particular gauge is a combination vacuum and pressure gauge used for automotive diagnosis:

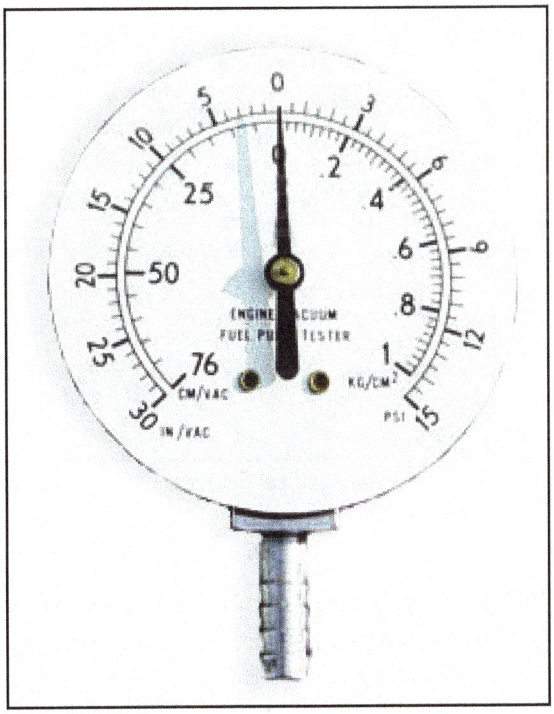

Indicator side with card and dial.

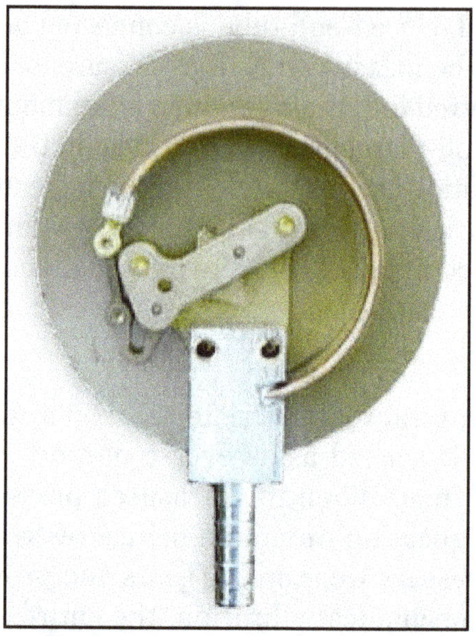

Mechanical side with Bourdon tube.

- the left side of the face, used for measuring manifold vacuum, is calibrated in centimetres of mercury on its inner scale and inches of mercury on its outer scale.
- the right portion of the face is used to measure fuel pump pressure or turbo boost and is calibrated in fractions of 1 kgf/cm^2 on its inner scale and pounds per square inch on its outer scale.

Mechanical Details

Mechanical details.

Stationary parts:

- A: Receiver block. This joins the inlet pipe to the fixed end of the Bourdon tube (1) and secures the chassis plate (B). The two holes receive screws that secure the case.
- B: Chassis plate. The face card is attached to this. It contains bearing holes for the axles.
- C: Secondary chassis plate. It supports the outer ends of the axles.

- D: Posts to join and space the two chassis plates.

Moving Parts:

1. Stationary end of Bourdon tube. This communicates with the inlet pipe through the receiver block.
2. Moving end of Bourdon tube. This end is sealed.
3. Pivot and pivot pin.
4. Link joining pivot pin to lever (5) with pins to allow joint rotation.
5. Lever. This is an extension of the sector gear (7).
6. Sector gear axle pin.
7. Sector gear.
8. Indicator needle axle. This has a spur gear that engages the sector gear (7) and extends through the face to drive the indicator needle. Due to the short distance between the lever arm link boss and the pivot pin and the difference between the effective radius of the sector gear and that of the spur gear, any motion of the Bourdon tube is greatly amplified. A small motion of the tube results in a large motion of the indicator needle.
9. Hair spring to preload the gear train to eliminate gear lash and hysteresis.

Diaphragm

A second type of aneroid gauge uses deflection of a flexible membrane that separates regions of different pressure. The amount of deflection is repeatable for known pressures so the pressure can be determined by using calibration. The deformation of a thin diaphragm is dependent on the difference in pressure between its two faces. The reference face can be open to atmosphere to measure gauge pressure, open to a second port to measure differential pressure, or can be sealed against a vacuum or other fixed reference pressure to measure absolute pressure. The deformation can be measured using mechanical, optical or capacitive techniques. Ceramic and metallic diaphragms are used.

Useful range: above 10^{-2} Torr (roughly 1 Pa)

For absolute measurements, welded pressure capsules with diaphragms on either side are often used.

shape:

- Flat
- corrugated
- flattened tube
- capsule

Bellows

A pile of pressure capsules with corrugated diaphragms in an aneroid barograph.

In gauges intended to sense small pressures or pressure differences, or require that an absolute pressure be measured, the gear train and needle may be driven by an enclosed and sealed bellows chamber, called an aneroid, which means "without liquid". (Early barometers used a column of liquid such as water or the liquid metal mercury suspended by a vacuum.) This bellows configuration is used in aneroid barometers (barometers with an indicating needle and dial card), altimeters, altitude recording barographs, and the altitude telemetry instruments used in weather balloon radiosondes. These devices use the sealed chamber as a reference pressure and are driven by the external pressure. Other sensitive aircraft instruments such as air speed indicators and rate of climb indicators (variometers) have connections both to the internal part of the aneroid chamber and to an external enclosing chamber.

Magnetic Coupling

These gauges use the attraction of two magnets to translate differential pressure into motion of a dial pointer. As differential pressure increases, a magnet attached to either a piston or rubber diaphragm moves. A rotary magnet that is attached to a pointer then moves in unison. To create different pressure ranges, the spring rate can be increased or decreased.

Spinning Rotor Gauge

The spinning rotor gauge works by measuring the amount a rotating ball is slowed by the viscosity of the gas being measured. The ball is made of steel and is magnetically levitated inside a steel tube closed at one end and exposed to the gas to be measured at the other. The ball is brought up to speed (about 2500 rad/s), and the speed measured after switching off the drive, by electromagnetic transducers. The range of the instrument is 10^{-5} to 10^{2} Pa (10^{3} Pa with less accuracy). It is accurate and stable enough to be used as a secondary standard. The instrument requires some skill and knowledge to use correctly. Various corrections must be applied and the ball must be spun at a pressure well below the intended measurement pressure for five hours before using. It is most useful in calibration and research laboratories where high accuracy is required and qualified technicians are available.

Electronic Pressure Sensors

Piezoresistive Strain Gage:

Uses the piezoresistive effect of bonded or formed strain gauges to detect strain due to applied pressure.

Capacitive:

Uses a diaphragm and pressure cavity to create a variable capacitor to detect strain due to applied pressure.

Magnetic:

Measures the displacement of a diaphragm by means of changes in inductance (reluctance), LVDT, Hall Effect, or by eddy current principle.

Piezoelectric:

Uses the piezoelectric effect in certain materials such as quartz to measure the strain upon the sensing mechanism due to pressure.

Optical:

Uses the physical change of an optical fiber to detect strain due to applied pressure.

Potentiometric:

Uses the motion of a wiper along a resistive mechanism to detect the strain caused by applied pressure.

Resonant:

Uses the changes in resonant frequency in a sensing mechanism to measure stress, or changes in gas density, caused by applied pressure.

Thermal Conductivity

Generally, as a real gas increases in density -which may indicate an increase in pressure- its ability to conduct heat increases. In this type of gauge, a wire filament is heated by running current through it. A thermocouple or resistance thermometer (RTD) can then be used to measure the temperature of the filament. This temperature is dependent on the rate at which the filament loses heat to the surrounding gas, and therefore on the thermal conductivity. A common variant is the Pirani gauge, which uses a single platinum filament as both the heated element and RTD. These gauges are accurate from 10^{-3} Torr to 10 Torr, but their calibration is sensitive to the chemical composition of the gases being measured.

Pirani (One Wire)

A Pirani gauge consist of a metal wire open to the pressure being measured. The wire is heated by a current flowing through it and cooled by the gas surrounding it. If the gas pressure is reduced, the cooling effect will decrease, hence the equilibrium temperature of the wire will increase. The resistance of the wire is a function of its temperature: by measuring the voltage across the wire and the current flowing through it, the resistance (and so the gas pressure) can be determined. This type of gauge was invented by Marcello Pirani.

Two-wire

In two-wire gauges, one wire coil is used as a heater, and the other is used to measure temperature due to convection. Thermocouple gauges and thermistor gauges work in this manner using thermocouple or thermistor, respectively, to measure the temperature of the heated wire.

Ionization Gauge

Ionization gauges are the most sensitive gauges for very low pressures (also referred to as hard or high vacuum). They sense pressure indirectly by measuring the electrical ions produced when the gas is bombarded with electrons. Fewer ions will be produced by lower density gases. The calibration of an ion gauge is unstable and dependent on the nature of the gases being measured, which is not always known. They can be calibrated against a McLeod gauge which is much more stable and independent of gas chemistry.

Thermionic emission generate electrons, which collide with gas atoms and generate positive ions. The ions are attracted to a suitably biased electrode known as the collector. The current in the collector is proportional to the rate of ionization, which is a function of the pressure in the system. Hence, measuring the collector current gives the gas pressure. There are several sub-types of ionization gauge.

Useful range: 10^{-10} - 10^{-3} torr (roughly 10^{-8} - 10^{-1} Pa)

Most ion gauges come in two types: hot cathode and cold cathode. In the hot cathode version, an electrically heated filament produces an electron beam. The electrons travel through the gauge and ionize gas molecules around them. The resulting ions are collected at a negative electrode. The current depends on the number of ions, which depends on the pressure in the gauge. Hot cathode gauges are accurate from 10^{-3} Torr to 10^{-10} Torr. The principle behind cold cathode version is the same, except that electrons are produced in the discharge of a high voltage. Cold Cathode gauges are accurate from 10^{-2} Torr to 10^{-9} Torr. Ionization gauge calibration is very sensitive to construction geometry, chemical composition of gases being measured, corrosion and surface deposits. Their calibration can be invalidated by activation at atmospheric pressure or low vacuum. The composition of gases at high vacuums will usually be unpredictable, so a mass spectrometer must be used in conjunction with the ionization gauge for accurate measurement.

Hot Cathode

A hot-cathode ionization gauge is composed mainly of three electrodes acting together as a triode, wherein the cathode is the filament. The three electrodes are a collector or plate, a filament, and a grid. The collector current is measured in picoamps by an electrometer. The filament voltage to ground is usually at a potential of 30 volts, while the grid voltage at 180–210 volts DC, unless there is an optional electron bombardment feature, by heating the grid, which may have a high potential of approximately 565 volts. The most common ion gauge is the hot-cathode Bayard-Alpert gauge, with a small ion collector inside the grid. A glass envelope with an opening to the vacuum can surround the electrodes, but usually the Nude Gauge is inserted in the vacuum chamber directly, the pins being fed through a ceramic plate in the wall of the chamber. Hot-cathode gauges can be damaged or lose their calibration if they are exposed to atmospheric pressure or even low vacuum while hot. The measurements of a hot-cathode ionization gauge are always logarithmic.

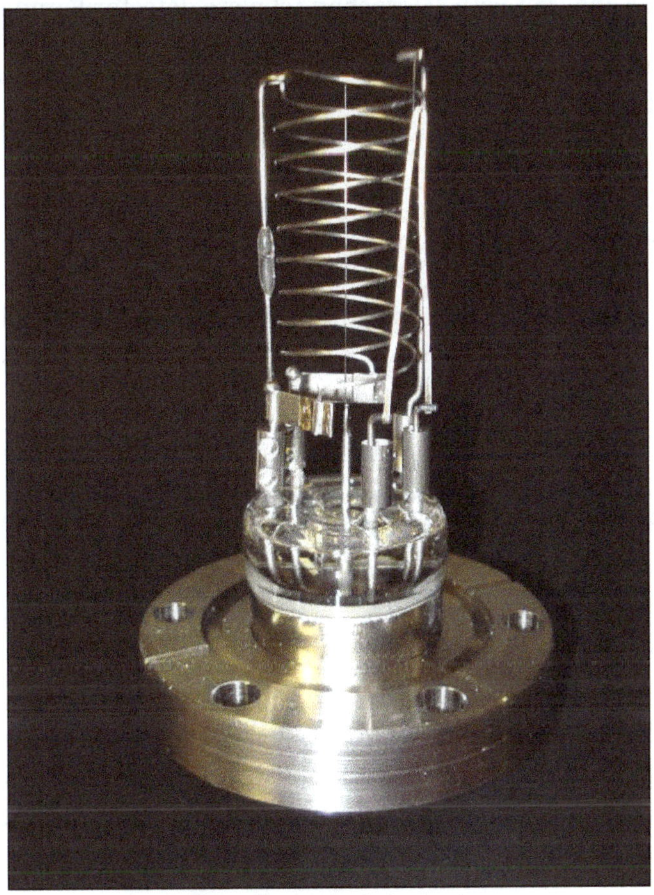

Bayard-Alpert hot-cathode ionization gauge.

Electrons emitted from the filament move several times in back and forth movements around the grid before finally entering the grid. During these movements, some electrons collide with a gaseous molecule to form a pair of an ion and an electron (Electron ionization). The number of these ions is proportional to the gaseous molecule density multiplied by the electron current emitted from the filament, and these ions pour into the collector to form an ion current. Since the gaseous molecule density is proportional to the pressure, the pressure is estimated by measuring the ion current.

The low-pressure sensitivity of hot-cathode gauges is limited by the photoelectric effect. Electrons hitting the grid produce x-rays that produce photoelectric noise in the ion collector. This limits the range of older hot-cathode gauges to 10^{-8} Torr and the Bayard-Alpert to about 10^{-10} Torr. Additional wires at cathode potential in the line of sight between the ion collector and the grid prevent this effect. In the extraction type the ions are not attracted by a wire, but by an open cone. As the ions cannot decide which part of the cone to hit, they pass through the hole and form an ion beam. This ion beam can be passed on to a:

- Faraday cup.
- Microchannel plate detector with Faraday cup.
- Quadrupole mass analyzer with Faraday cup.
- Quadrupole mass analyzer with Microchannel plate detector Faraday cup.
- ion lens and acceleration voltage and directed at a target to form a sputter gun. In this case a valve lets gas into the grid-cage.

Cold Cathode

There are two subtypes of cold-cathode ionization gauges: the Penning gauge (invented by Frans Michel Penning), and the Inverted magnetron, also called a Redhead gauge. The major difference between the two is the position of the anode with respect to the cathode. Neither has a filament, and each may require a DC potential of about 4 kV for operation. Inverted magnetrons can measure down to $1x10^{-12}$ Torr.

Likewise, cold-cathode gauges may be reluctant to start at very low pressures, in that the near-absence of a gas makes it difficult to establish an electrode current - in particular in Penning gauges, which use an axially symmetric magnetic field to create path lengths for electrons that are of the order of metres. In ambient air, suitable ion-pairs are ubiquitously formed by cosmic radiation; in a Penning gauge, design features are used to ease the set-up of a discharge path. For example, the electrode of a Penning gauge is usually finely tapered to facilitate the field emission of electrons.

Maintenance cycles of cold cathode gauges are, in general, measured in years, depending on the gas type and pressure that they are operated in. Using a cold cathode gauge in gases with substantial organic components, such as pump oil fractions, can result in the growth of delicate carbon films and shards within the gauge that eventually either short-circuit the electrodes of the gauge or impede the generation of a discharge path.

Dynamic Transients

When fluid flows are not in equilibrium, local pressures may be higher or lower than the average pressure in a medium. These disturbances propagate from their source as longitudinal pressure variations along the path of propagation. This is also called sound. Sound pressure is the instantaneous local pressure deviation from the average pressure caused by a sound wave. Sound pressure can be measured using a microphone in air and a hydrophone in water. The effective sound pressure is the root mean square of the instantaneous sound pressure over a given interval of time. Sound pressures are normally small and are often expressed in units of microbar.

- frequency response of pressure sensors,
- resonance.

Calibration and Standards

Dead-weight tester. This uses known calibrated weights on a piston to generate a known pressure.

The American Society of Mechanical Engineers (ASME) has developed two separate and distinct standards on pressure Measurement, B40.100 and PTC 19.2. B40.100 provides guidelines on Pressure Indicated Dial Type and Pressure Digital Indicating Gauges, Diaphragm Seals, Snubbers, and Pressure Limiter Valves. PTC 19.2 provides instructions and guidance for the accurate determination of pressure values in support of the ASME Performance Test Codes. The choice of method, instruments, required calculations, and corrections to be applied depends on the purpose of the measurement, the allowable uncertainty, and the characteristics of the equipment being tested.

The methods for pressure measurement and the protocols used for data transmission are also provided. Guidance is given for setting up the instrumentation and determining the uncertainty of the measurement. Information regarding the instrument type, design, applicable pressure range, accuracy, output, and relative cost is provided. Information is also provided on pressure-measuring devices that are used in field environments i.e., Piston Gauges, Manometers, and Low-Absolute-Pressure (Vacuum) Instruments.

These methods are designed to assist in the evaluation of measurement uncertainty based on current technology and engineering knowledge, taking into account published instrumentation specifications and measurement and application techniques. This Supplement provides guidance in the use of methods to establish the pressure-measurement uncertainty.

European (CEN) Standard

- EN 472 : Pressure gauge - Vocabulary.
- EN 837-1 : Pressure gauges. Bourdon tube pressure gauges. Dimensions, metrology, requirements and testing.
- EN 837-2 : Pressure gauges. Selection and installation recommendations for pressure gauges.
- EN 837-3 : Pressure gauges. Diaphragm and capsule pressure gauges. Dimensions, metrology, requirements, and testing.

US ASME Standards

- B40.100-2013: Pressure gauges and Gauge attachments.
- PTC 19.2-2010 : Performance test code for pressure measurement.

Pressure measurement is important in many fluid mechanics related applications. From appropriate pressure measurements velocity, aerodynamic forces and moments can be determined. Pressure is measured by the force acting on unit area. Measuring devices usually indicate differential pressure i.e. in relation with atmospheric pressure. This is called gauge pressure. The measured pressure may be positive or negative with reference to the atmospheric pressure .A negative gauge pressure is referred to as vacuum.

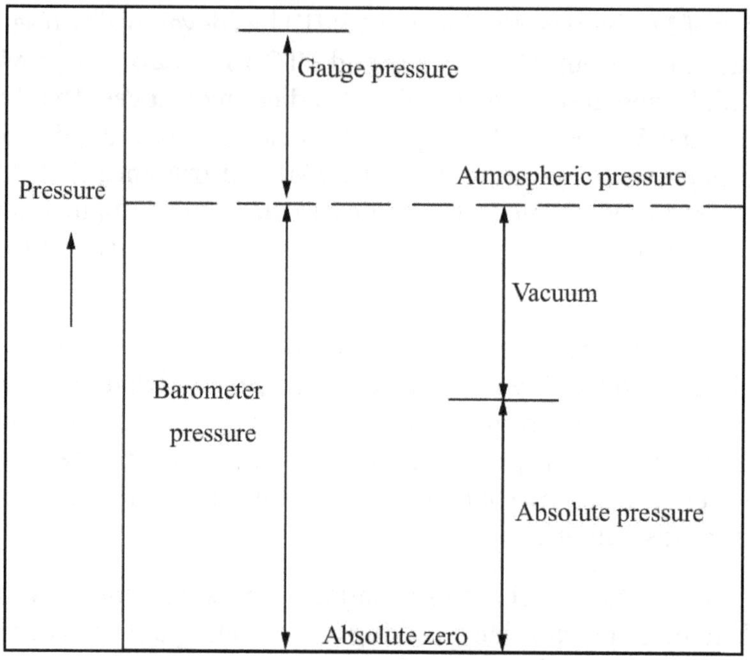

Explanation of the pressure terminology.

Units of Pressure

1Pascal(1N/m^2)	= 10dyne/cm^2
1mmHg	= 133.32pascals
	= 13.595mm Water
Standard atmosphere	= 1.013 * 10^{5}N/m^2
1 millibar	= 1000 dyn/cm^2
1 micron	= 10^{-6}mHg
1 torr	= 1 mmHg
	= 1000micron

Absolute Pressure

Absolute pressure is determined as algebraic sum of the readings of a barometer and of a manometer showing the gauge pressure. Manometers which measure absolute pressure are also available. They measure the pressure with reference to absolute zero pressure.

Pressure Measuring Devices

Main characteristics of manometers are pressure range, accuracy, sensitivity and speed of response. Pressure range of manometers varies from almost perfect vacuum to several hundreds of atmosphere. The conventional instruments used for pressure measurement are divided into the following groups:

1. Liquid column manometers
2. Pressure gauges with elastic sensing elements
3. Pressure transducers
4. Manometers for low absolute pressures
5. Manometers for very high absolute pressures

Liquid Column Manometers

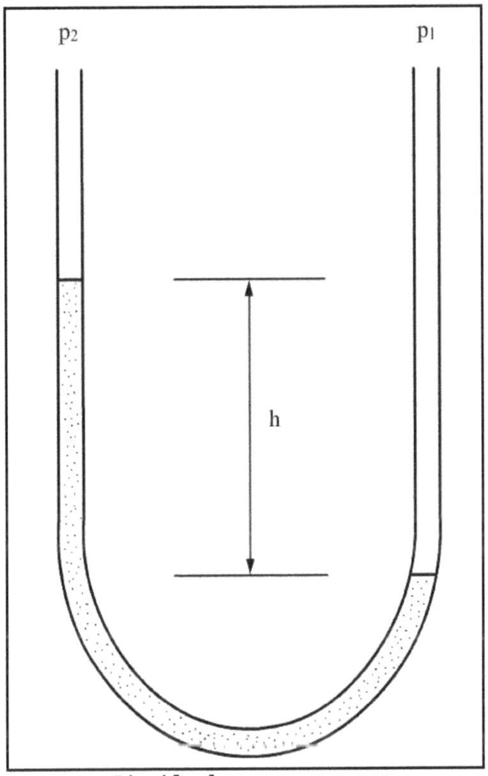

Liquid column manometer.

For amplifying the deflection in a liquid column manometer, liquids with lower density could be used or one of the limbs of the manometer may be inclined. Commonly used manometric liquids are mercury, water or alcohol. Some of the important and desirable properties of the manometric liquids are:

- High chemical stability
- Low viscosity
- Low capillary constant
- Low coefficient of thermal expansion
- Low volatility
- Low vapour pressure

High thermal stability and low volatility are important for maintaining a constant. Specific gravity. High viscosity causes transmission lags .Thermal expansion causes changes in zero reading. While measuring low pressures, vapour pressure of the manometric fluid is an important consideration. Properties of some of the commonly used fluids are given in the table below.

Table: Typical properties of manometric fluids.

Fluid	Specific gravity	B.P.(^{o}C) at 760mm Hg	Surface Tension dyn/cm	Viscosity Cp	Coeft.of Volumetric Expansion $*10^5$
Methyl Alcohol	0.792	64.7	22.6	0.59	
Ethyl Alcohol	0.789	78.4	22.0	1.9	110
Mercury	13.55	356.59	465	1.55	18
Toluene	0.866	110.8	28.4		
CCl_4	1.594	76.8	26.8	0.97	

Inclined Manometer

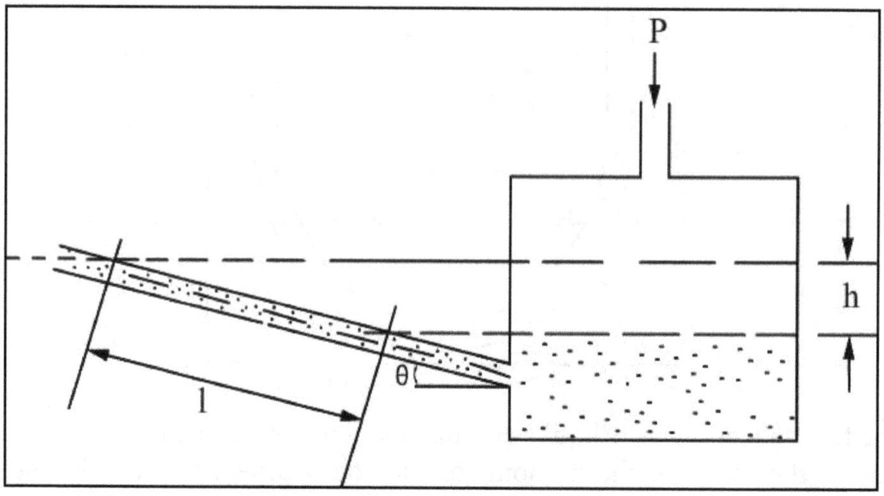

Inclined manometer.

The figure above shows the amplification of the reading of applied pressure 'h' as

$$\frac{h}{l} = \sin\theta$$

Mercury Barometer

Barometer is the device used to measure the atmospheric pressure. Mercury barometer consists essentially of a glass tube sealed at one end and mounted vertically in a bowl or cistern of mercury so that the open end of the tube is submerged below the surface of mercury in the cistern. In the figure below (i) the zero level of mercury in the cistern is shown under the influence of atmospheric pressure p_1. When $p_2=p_1$ = atmospheric pressure, the zero level in the cistern can be marked. If the tube is open and if different pressures act in the cistern and in the tube, then there will be difference in the levels of mercury. If p_1 is greater than p_2 as shown in the figure below (ii), then mercury will be forced down in the cistern and corresponding rise will be there in the tube.

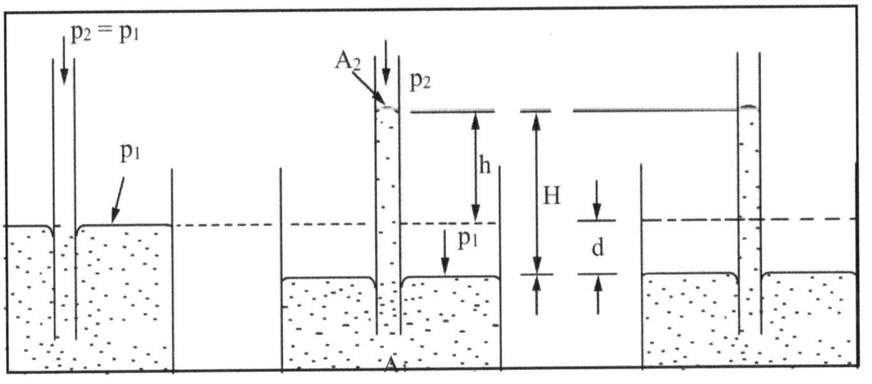

(i) (ii) (iii)

Principle of mercury barometer.

The balance of pressures will be given by $p_1 = p_2 + H\rho$ 1

(where H is the difference between levels of mercury in the cistern and in the tube and ρ the density of mercury). From the figure above (ii), H=h+d

The quantity of mercury that has left the cistern is same as that has risen in the tube.

$A_1 d = A_2 h$ where A_1 and A_2 are areas of the cistern and the tube respectively.

$$d = \frac{A_2}{A_1} h$$

$$\therefore H = h + h\left(\frac{A_2}{A_1}\right)$$

$$= h\left(1 + \frac{A_2}{A_1}\right) \quad \text{..............................} 2$$

Replacing H in equation 1

$$p_1 = p_2 + h\left(1 + \frac{A_2}{A_1}\right)\rho \quad \text{.....................} 3$$

$p_2 \to 0$ in practical cases as the tube will be evacuated and sealed.

$$p_1 = h\left(1 + \frac{A_2}{A_1}\right)\rho \quad \text{.....................................} 4$$

Micromanometer

For accurate measurement of extremely small pressure differences micromanometers are used. In the figure given below, the instrument is initially adjusted such that $p_1 = p_2$.

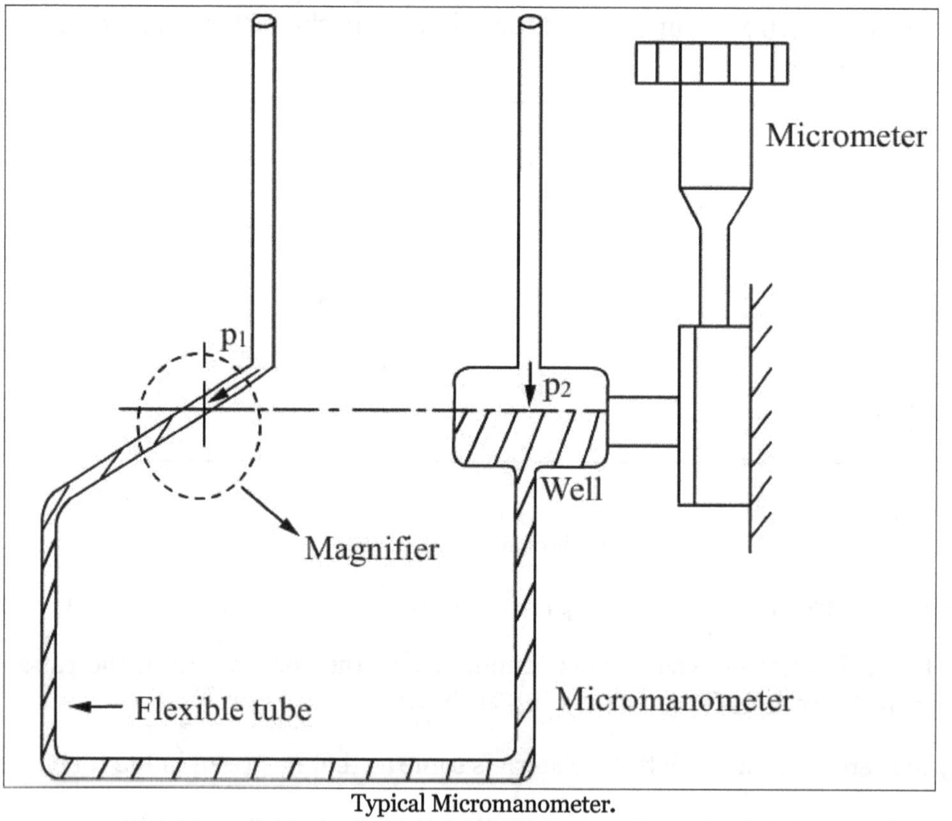

Typical Micromanometer.

The meniscus in the inclined tube is located at a reference level fixed by the hairline viewed through the magnifier. The reading of the micrometer is noted. Application of the unknown pressure difference causes the meniscus to move off the hairline but it can be restored to the initial position by raising or lowering the well (mercury sump) The difference in the initial and final micrometer readings gives the height of the mercury column and hence the pressure. Pressures as low as 0.025mm water column can be measured.

Mechanical Manometers

Mechanical manometers provide faster response than liquid column manometers. In liquid column measurements, lag is due to the displacements of the liquid. In elastic sensing element type of manometers the time lag is due to the time required for equalisation of pressure to be measured with that in the sensing chamber. The deformation of elastic sensing elements is measured with the aid of kinematic, optical or electrical systems. There are three types of elastic sensing elements which are (i) Bourdon Tubes (ii) Diaphragms (flat or corrugated) (iii) Bellows

Bourdon Tube

Bourdon tube is the oldest pressure sensing element. It is a length of metal tube of elliptical cross section and shaped into letter 'C'.

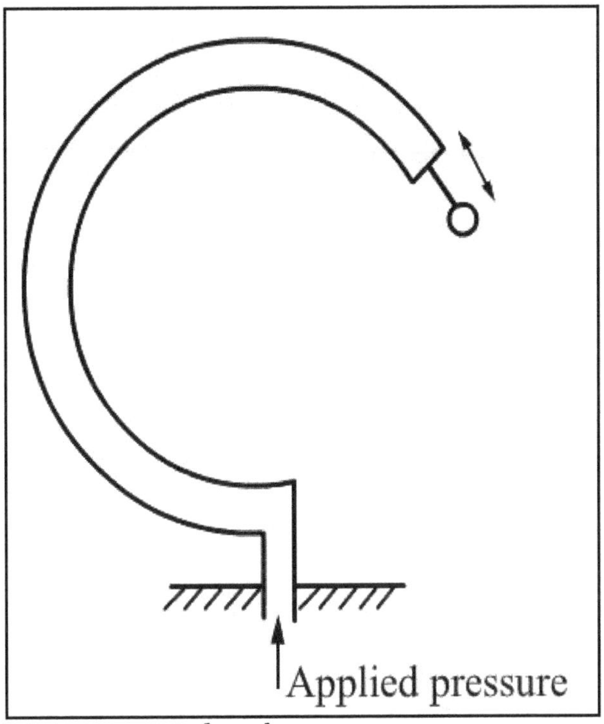

Bourdon tube pressure gauge.

A tube of elliptical cross section has a smaller volume than a circular one of the same length and perimeter. When connected to the pressure source it is made to accommodate more of the fluid. Resultant of all reactions will produce maximum displacement at the free end. Within close limits, the change in angle subtended at the centre by a tube is proportional to the change of internal pressure and within the limits of proportionality of the material; the displacement of the free end is proportional to the applied pressure.

The ratio between major and minor axes decides the sensitivity of the Bourdon tube. The larger is the higher is the sensitivity. Materials of the Bourdon tube is Phosphor bronze, Beryllium bronze or Beryllium Copper.

Elastic Diaphragms

The pressure sensors making use of elastic diaphragms consist of a diaphragm fixed in a tubular member. The pressure to be measured is applied on one side. The mathematical relation between pressure and central deflection for a flat circular diaphragm is given by

$$p = \frac{16Et^4}{3a^4\left(1-\mu^2\right)}\left[\frac{y_c}{t} + 0.488\left(\frac{y_c}{t}\right)^3 + \ldots\right] \quad \ldots\ldots\ldots\ldots\ldots 5$$

To have a linear pressure deflection relation, the second and later terms should be small.

p = applied pressure

t = thickness

a = radius

yc = central deflection

E = Youngs modulus

μ = Poisson's ratio $\rightarrow \dfrac{\textit{lateral strain}}{\textit{Axial strain}}$

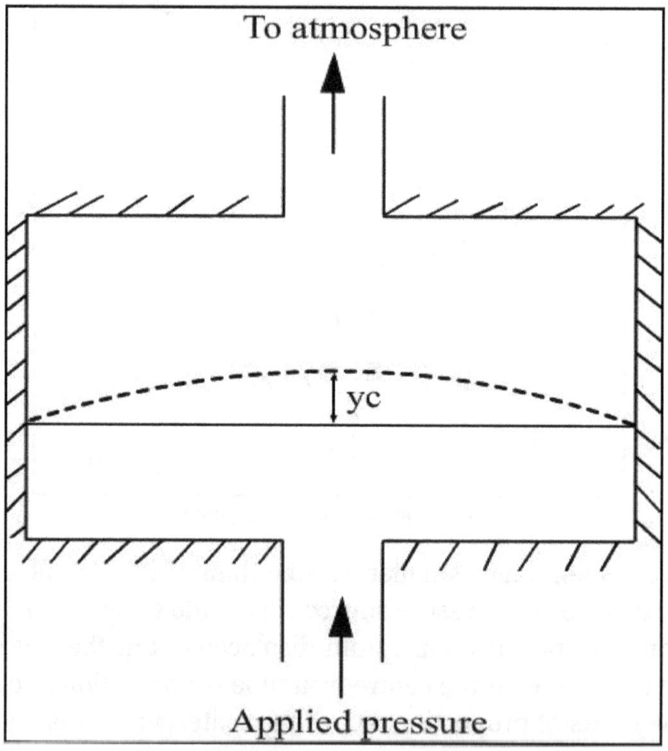

Diaphragm pressure gauge.

As seen from the equation the pressure deflection relation is not linear. If a non-linearity of 5% is acceptable the deflection must be less than 1/3 of thickness. Neglecting the higher order terms within brackets,

$$p = \frac{16Et^4}{3a^4\Pi(1-\mu^2)} \times \frac{y_c}{t}$$

$$p = \frac{16Et^3 y_c}{3a^4(1-\mu^2)}$$

$$\frac{y_c}{p} = \frac{\cancel{y_c}\, 3a^4(1\text{-}\mu^2)}{16Et^3 \cancel{y_c}} \quad \ldots\ldots\ldots\ldots\ldots\ldots\ldots\ldots 6$$

$$\frac{y_c}{p} = K = \frac{3(1-\mu^2)a^4}{16E\,t^3} \quad \ldots\ldots\ldots\ldots\ldots\ldots\ldots\ldots 7$$

where 'K' is the sensitivity of the device.

The natural frequency of the diaphragm may be written as follows:

$$\omega = 10.2\left[\frac{E}{12p_d\left(1-\mu^2\right)}\right]^{1/2}\frac{t}{a^2}\,rad/s \qquad ....8$$

Diaphragm has infinite number of natural frequencies and the lowest of them is considered here. From the two equations for sensitivity 'K' and the natural frequency ω.

Sensitivity (K) and natural frequency (ω) are related by

$$\frac{y_c}{p} = K = \frac{1.63}{\omega^2 t \rho_d} \qquad ....9$$

(ρ_d is density of diaphragm material)

Sensitivity is inversely proportional to square of natural frequency. Sensitivity can be improved by lowering the natural frequency. The natural frequency should be 3 to 4 times the frequency of pressure pulsations. Compared to bourdon tubes diaphragms have higher natural frequencies. Hence, they can be used for the measurement of dynamic pressures of higher frequency fluctuations. They can be installed flush with the surface of the body and hence there is only much less transmission lag.

Corrugated Diaphragms

Corrugated diaphragms permit considerably larger deflections than flat diaphragms. Their number and depth control the response and sensitivity. The greater the number and depth, the more linear is its deflection and greater is the sensitivity.

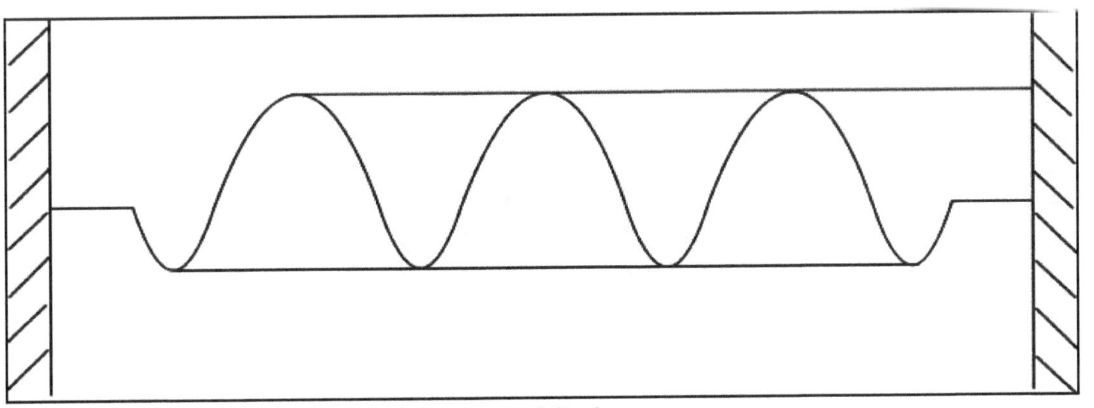

Corrugated diaphragm.

Capsules, Bellows

For even larger deflections than the diaphragms corrugated diaphragms are made in to boxes or bellows. Bellows are most commonly used to measure small steady pressures.

Pressure Sensor

Digital air pressure sensor.

Miniature digital barometric pressure sensor.

Some pressure sensors, such as those found in some traffic enforcement cameras, function in a binary (off/on) manner, i.e., when pressure is applied to a pressure sensor, the sensor acts to complete or break an electrical circuit. These types of sensors are also known as a pressure switch.

Types of Pressure Measurements

silicon piezoresistive pressure sensors.

Pressure sensors can be classified in terms of pressure ranges they measure, temperature ranges of operation, and most importantly the type of pressure they measure. Pressure sensors are variously named according to their purpose, but the same technology may be used under different names.

- Absolute pressure sensor

This sensor measures the pressure relative to perfect vacuum.

- Gauge pressure sensor

This sensor measures the pressure relative to atmospheric pressure. A tire pressure gauge is an example of gauge pressure measurement; when it indicates zero, then the pressure it is measuring is the same as the ambient pressure.

- Vacuum pressure sensor

This term can cause confusion. It may be used to describe a sensor that measures pressures below atmospheric pressure, showing the difference between that low pressure and atmospheric pressure (i.e. negative gauge pressure), but it may also be used to describe a sensor that measures low pressure relative to perfect vacuum (i.e. absolute pressure).

- Differential pressure sensor

This sensor measures the difference between two pressures, one connected to each side of the sensor. Differential pressure sensors are used to measure many properties, such as pressure drops across oil filters or air filters, fluid levels (by comparing the pressure above and below the liquid) or flow rates (by measuring the change in pressure across a restriction). Technically speaking, most pressure sensors are really differential pressure sensors; for example a gauge pressure sensor is merely a differential pressure sensor in which one side is open to the ambient atmosphere.

- Sealed pressure sensor

This sensor is similar to a gauge pressure sensor except that it measures pressure relative to some fixed pressure rather than the ambient atmospheric pressure (which varies according to the location and the weather).

Pressure-sensing Technology

There are two basic categories of analog pressure sensors,

Force collector types These types of electronic pressure sensors generally use a force collector (such a diaphragm, piston, bourdon tube, or bellows) to measure strain (or deflection) due to applied force over an area (pressure).

- *Piezoresistive strain gauge*

  Uses the piezoresistive effect of bonded or formed strain gauges to detect strain due to applied pressure, resistance increasing as pressure deforms the material. Common technology types are Silicon (Monocrystalline), Polysilicon Thin Film, Bonded Metal Foil, Thick Film, and Sputtered Thin Film. Generally, the strain gauges are connected to form a Wheatstone bridge circuit to maximize the output of the sensor and to reduce sensitivity to errors. This is the most commonly employed sensing technology for general purpose pressure measurement.

- *Capacitive*

  Uses a diaphragm and pressure cavity to create a variable capacitor to detect strain due to applied pressure, capacitance decreasing as pressure deforms the diaphragm. Common technologies use metal, ceramic, and silicon diaphragms.

- *Electromagnetic*

  Measures the displacement of a diaphragm by means of changes in inductance (reluctance), LVDT, Hall Effect, or by eddy current principle.

- *Piezoelectric*

  Uses the piezoelectric effect in certain materials such as quartz to measure the strain upon the sensing mechanism due to pressure. This technology is commonly employed for the measurement of highly dynamic pressures.

- *Optical*

  Techniques include the use of the physical change of an optical fiber to detect strain due to applied pressure. A common example of this type utilizes Fiber Bragg Gratings. This technology is employed in challenging applications where the measurement may be highly remote, under high temperature, or may benefit from technologies inherently immune to electromagnetic interference. Another analogous technique utilizes an elastic film constructed in layers that can change reflected wavelengths according to the applied pressure (strain).

- *Potentiometric*

  Uses the motion of a wiper along a resistive mechanism to detect the strain caused by applied pressure.

Other Types

These types of electronic pressure sensors use other properties (such as density) to infer pressure of a gas, or liquid.

- *Resonant*

  Uses the changes in resonant frequency in a sensing mechanism to measure stress, or changes in gas density, caused by applied pressure. This technology may be used in conjunction with a force collector, such as those in the category above. Alternatively, resonant technology may be employed by exposing the resonating element itself to the media, whereby the resonant frequency is dependent upon the density of the media. Sensors have been made out of vibrating wire, vibrating cylinders, quartz, and silicon MEMS. Generally, this technology is considered to provide very stable readings over time.

- *Thermal*

  Uses the changes in thermal conductivity of a gas due to density changes to measure pressure. A common example of this type is the Pirani gauge.

- *Ionization*

  Measures the flow of charged gas particles (ions) which varies due to density changes to measure pressure. Common examples are the Hot and Cold Cathode gauges.

Applications

There are many applications for pressure sensors:

- Pressure sensing

This is where the measurement of interest is pressure, expressed as a force per unit area. This is useful in weather instrumentation, aircraft, automobiles, and any other machinery that has pressure functionality implemented.

- Altitude sensing

This is useful in aircraft, rockets, satellites, weather balloons, and many other applications. All these applications make use of the relationship between changes in pressure relative to the altitude. This relationship is governed by the following equation:

$$h = (1 - (P / P_{\text{ref}})^{0.190284}) \times 145366.45\text{ft}$$

This equation is calibrated for an altimeter, up to 36,090 feet (11,000 m). Outside that range, an error will be introduced which can be calculated differently for each different pressure sensor. These error calculations will factor in the error introduced by the change in temperature as we go up.

Barometric pressure sensors can have an altitude resolution of less than 1 meter, which is significantly better than GPS systems (about 20 meters altitude resolution). In navigation applications altimeters are used to distinguish between stacked road levels for car navigation and floor levels in buildings for pedestrian navigation.

- Flow sensing

This is the use of pressure sensors in conjunction with the venturi effect to measure flow. Differential pressure is measured between two segments of a venturi tube that have a different aperture. The pressure difference between the two segments is directly proportional to the flow rate through the venturi tube. A low pressure sensor is almost always required as the pressure difference is relatively small.

- Level / depth sensing

A pressure sensor may also be used to calculate the level of a fluid. This technique is commonly employed to measure the depth of a submerged body (such as a diver or submarine), or level of contents in a tank (such as in a water tower). For most practical purposes, fluid level is directly proportional to pressure. In the case of fresh water where the contents are under atmospheric pressure, 1psi = 27.7 inH2O / 1Pa = 9.81 mmH2O. The basic equation for such a measurement is;

$$P = \rho g h$$

where P = pressure, ρ = density of the fluid, g = standard gravity, h = height of fluid column above pressure sensor

- Leak testing

A pressure sensor may be used to sense the decay of pressure due to a system leak. This is commonly done by either comparison to a known leak using differential pressure, or by means of utilizing the pressure sensor to measure pressure change over time.

Ratiometric Correction of Transducer Output

Piezoresistive transducers configured as Wheatstone bridges often exhibit ratiometric behavior with respect not only to the measured pressure, but also the transducer supply voltage.

$$V_{\text{out}} = \frac{P \times K \times Vs_{\text{actual}}}{Vs_{\text{ideal}}}$$

where:

V_{out} is the output voltage of the transducer.

P is the actual measured pressure.

K is the nominal transducer scale factor (given an ideal transducer supply voltage) in units of voltage per pressure.

Vs_{actual} is the actual transducer supply voltage.

Vs_{ideal} is the ideal transducer supply voltage.

Correcting measurements from transducers exhibiting this behavior requires measuring the actual transducer supply voltage as well as the output voltage and applying the inverse transform of this behavior to the output signal:

$$P = \frac{V_{\text{out}} \times Vs_{\text{ideal}}}{K \times Vs_{\text{actual}}}$$

Note: Common mode signals often present in transducers configured as Wheatstone bridges are not considered in this analysis.

Pressure Transducers

Diaphragm Type Pressure Transducers

They convert the pressure to be measured into electrical signals. For example, pressure transducers whose operating principle is based on measuring changes in inductive, capacitive or ohmic resistances caused by the deformation of an elastic element.

Capacitance $C = \in_a a / d$.................................10

Where $\in_a \rightarrow$ absolute permittivity

$$\frac{\in_a}{\in_0} = K \frac{1}{4\Pi \times 9 \times 10^9 E / m}$$

a = area plates

d= distance between the plates

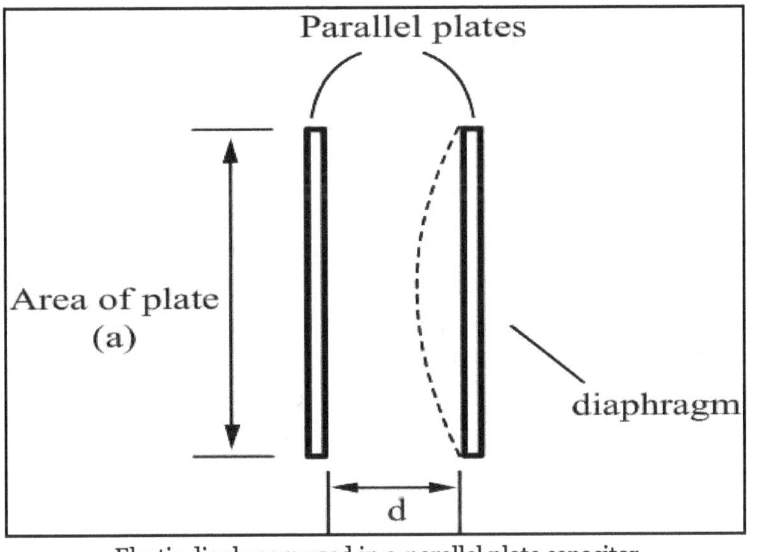

Elastic diaphragm used in a parallel plate capacitor.

With the application of pressure, the diaphragm deforms with the central deflection as shown. This changes the capacitance in an electrical circuit which can be calibrated against pressure to be measured.

Piezo-Electric Pressure Transducers

The word piezo-electric is derived from the Greek word "piezein" meaning squeeze or press. Certain materials possess the ability to generate electrical potential when subjected to mechanical strain. Conversely, they change dimensions when supplied with voltage. The potential developed by application of stress is not held under static conditions. Dynamic pressures in the range of frequencies from kHz to 100MHz can be measured using piezo- electric transducers. The use of piezo-electric effect is limited to dynamic measurements. Some materials exhibiting piezo electric properties are quartz, tourmaline, barium titanate, and Lead zirconate. Quartz is the preferred material, as it possesses good mechanical properties. Also, it is a good insulator least influenced by moisture.

Pressure Sensitive Paints (PSP)

They are composed of luminescent molecules dispersed in Oxygen permeable polymer binder. When PSP is exposed to blue or an ultraviolet light the luminescent molecules are excited to a higher energy level.

From this excited state they can discharge in three ways:

a) by discharging light.

b) by transferring energy to the polymer binder(heating).

c) colliding with Oxygen molecules.

Since the luminescent molecules react with Oxygen they collide and release light at the same time. The amount of light emitted is inversely proportional to the amount of Oxygen molecules on the surface.

Measurement of High Pressures

Electrical Resistance Gauges – Principle of Operation

Bourdon Tube or strain gauge can be used for high pressures. Very high pressures (say above 1000bars) may be measured by means of electrical resistance gauges which are known as Bridgeman gauge. By way of principle of operation, they make use of resistance change brought about by direct application of pressure to the conductor itself. Referring to the Figure, the sensing element is the thin wire of Magnanin (84 Cu + 12 Mn + 4 Ni) or an alloy of Gold and Chromium (2.1%) which is loosely wound. When pressure is applied, bulk compression effects produce a change in resistance which may be calibrated against pressure. The general relation between electrical and mechanical may be derived as follows:

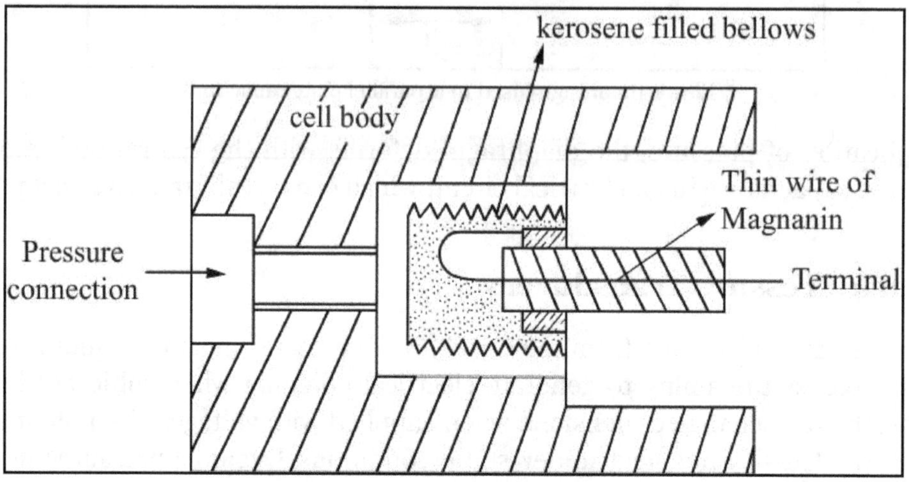

Sensor of a Bridgemans Gauge.

$$R = \frac{\rho L}{A} = \frac{\rho L}{CD^2} \qquad \text{......................................11}$$

R = Resistance, ohms

L = Length of conductors

A = Area = CD^2

where D is the diameter of the circular conductor and C a constant ρ = Resistivity ohm cm

If the conductor is strained each of the quantities in the equation will change.

Differentiating

$$dR = \frac{CD^2\left(Ld\rho + \rho dL\right) - 2C\rho DLdD}{\left(CD^2\right)^2}$$

$$= \frac{1}{CD^2}\left[\left(Ld\rho + \rho dL\right) - 2\rho L\frac{dD}{D}\right] \ldots\ldots\ldots\ldots\ldots\ldots 12$$

Dividing (12) by (11) gives

$$\frac{dR}{R} = \frac{dL}{L} - \frac{2dD}{D} + \frac{d\rho}{\rho}$$

The wire will be subjected to biaxial stresses because the ends in providing electrical continuity will not be subjected to pressure.

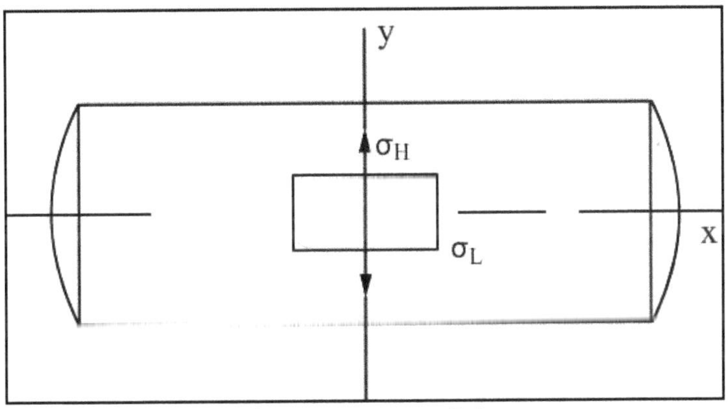

Bi-axial stresses on a small element.

Let us consider a general element subjected to stresses which may be denoted as σ_x *and* σ_y. Suppose σ_x *and* σ_y are applied one at a time. If σ_x is applied first, there will be strain in the

$$\text{x- direction} = \frac{\sigma_x}{E}$$

Because of Poisson's effect, the strain in y direction $= -\mu\frac{\sigma}{E}$

If σ_y is applied first

Strain in y direction $= \frac{\sigma_y}{E}$

Strain in x direction $= -\mu\frac{\sigma_y}{E}$

There will be a net strain of

$$\in_x = \frac{1}{E}(\sigma_x - \mu\sigma_y)$$

$$\in_y = -\frac{1}{E}(\sigma y - \mu\sigma_x)$$

Stresses are nothing but applied pressure

$\sigma_x = \sigma_y = -p$ assuming $\sigma_Z = 0$

Lateral Strain

$$\in_x = \in_y = \frac{dD}{D} = -\frac{p}{E}(1-\mu)$$

$$\in_Z = \frac{dL}{L} = 2\mu\frac{p}{E}$$

Biaxial stress situation is assumed.

Longitudinal Strain

$\therefore \frac{dR}{R} = \frac{dL}{L} - \frac{2dD}{D} + \frac{d\rho}{\rho}$ (Substitute from above)

$$= 2\mu\frac{p}{E} + 2\frac{p}{E}(1-\mu) + \frac{d\rho}{\rho}$$

$$= \frac{2p}{E} + \frac{d\rho}{\rho}$$

If specific resistance ρ does not depend on pressure $\frac{d\rho}{\rho}$ can be neglected.

$$\frac{dR}{R} = 2\frac{p}{E}$$

R = Ro (1 + bp) where, b = 2 / E.

$$\frac{dR/R}{p} = \frac{2}{E} + \frac{d\rho/\rho}{\rho} \quad \text{......................13}$$

where b is called pressure coefficient of resistance. Resistance varies linearly with pressure.

Ranges of Different Manometers

Liquid column manometers	10 – 0.5 x 10^6 pascals
Bourdon tubes	Vacuum to a few thousand bars
Diaphragms	Vacuum to a few hundred bars
Bellows	Vacuum to a hundred bar

Measurement of Vacuum

Pressures below atmosphere is vacuum. Very low pressures may be defined as below 1mmHg. Ultra low pressures is less than a milli micron (<10^{-3} micron)

Measurement of vacuum may be by two methods

Direct measurement

> Resulting in a displacement caused by the action of force [Spiral Bourdon tubes, flat or corrugat ed diaphragm, capsules and various other manometers]

Indirect measurement or inferential methods

> Pressure is determined through the measurement of certain pressure controlled properties such as volume, thermal conductivity etc.

McLeod Gauge

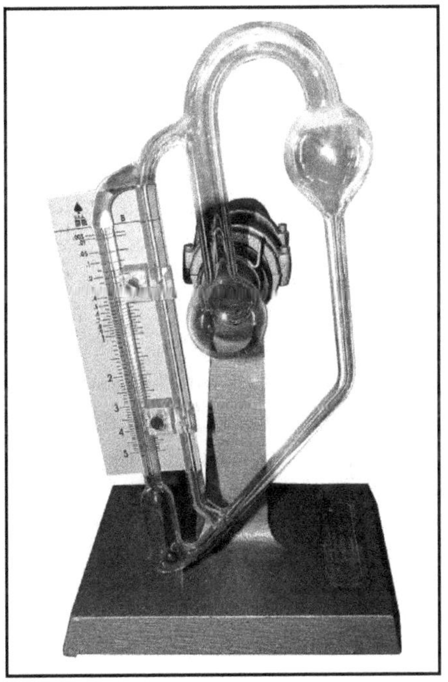

A glass McLeod gauge, drained of mercury.

A McLeod gauge is a scientific instrument used to measure very low pressures, down to 10^{-6} Torr. It was invented in 1874 by Herbert McLeod (1841–1923). McLeod gauges were once commonly found attached to equipment that operates under vacuum, such as a lyophilizer. Today, however, these gauges have largely been replaced by electronic vacuum gauges.

The design of a McLeod gauge is somewhat similar to a that of a mercury-column manometer. Typically it is filled with mercury. If used incorrectly, this mercury can escape and contaminate the vacuum system attached to the gauge.

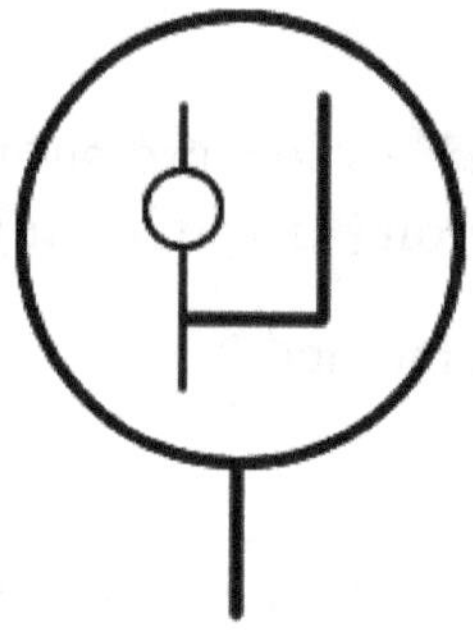

McLeod manometer symbol according to ISO 3753-1977 (E).

McLeod gauges operate by taking in a sample volume of gas from a vacuum chamber, then compressing it by tilting and infilling with mercury. The pressure in this smaller volume is then measured by a mercury manometer, and knowing the compression ratio (the ration of the initial and final volumes), the pressure of the original vacuum can be determined by applying Boyle's law.

This method is fairly accurate for non-condensible gases, such as oxygen and nitrogen. However, condensible gases, such as water vapour, ammonia, carbon dioxide, and pump-oil vapors may be in gaseous form in the low pressure of the vacuum chamber, but will condense when compressed by the McLeod gauge. The result is an erroneous reading, showing a pressure much lower than actually present. A cold trap may be used in conjunction with a McLeod gauge to condense these vapors before they enter the gauge.

The McLeod gauge has the advantage that it is simple to use and that its calibration is nearly the same for all non-condensable gases. The device can be manually operated and the scale read visually, or the process can be automated in various ways. For example, a small electric motor can periodically rotate the assembly to collect a gas sample. If a fine platinum wire is in the capillary tube, its resistance indicates the height of the mercury column around it.

Modern electronic vacuum gauges are simpler to use, less fragile, and do not present a mercury hazard, but their reading is highly dependent on the chemical nature of the gas being measured, and their calibration is unstable. For this reason, McLeod gauges continue to be used as a calibration standard for electronic gauges.

Inferential Methods

McLeod Gauge

The working of McLeod Gauge is based on Boyles' fundamental equation.

$$p_1 = \frac{p_2 V_2}{V_1}$$

where p and V refer to pressure and volume respectively and subscripts 1 and 2 refer to initial and final conditions. Conventional McLeod gauge is made of glass. Refer to the figure given below. It consists of the capillary 'C', bulb 'B' and the mercury sump which is connected to the lower end of the glass tube such that it can be moved up and down.

The pressure to be measured (the unknown pressure) is connected to the upper end of the glass part. When the mercury level in the gauge is below the cut off 'F', the unknown pressure fills the gauge including the bulb B and capillary C. When the mercury sump is moved up, the level in the gauge rises and when it reaches the cut off 'F' a known volume of gas at pressure to be measured is trapped in bulb B and capillary C.

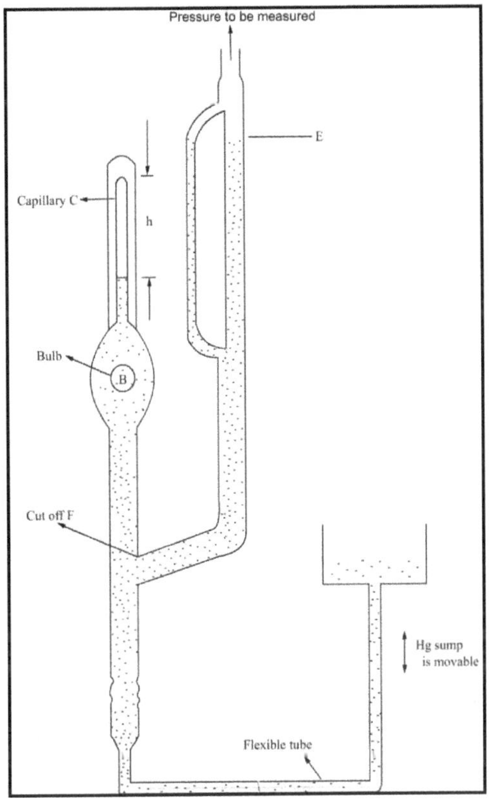

Mc Leod gauge.

Mercury is then forced up into the bulb and capillary. Assume the sump is raised to such a level that the gas at the pressure to be measured which filled the volume above the cut off is now compressed to the volume represented by the column h.

Suppose the original volume after then mercury reaches F is V_0. This is at a pressure being measured p_1

$$V_0 \times p_1 = ah \times (p_1 + h)$$

$$= p_1\ ah + ah^2$$

$$V_0 p_1 - p_1\ ah = ah^2$$

$$p_1 (V_0 - ah) = ah^2$$

As 'ah' is <<< Vo it is neglected.

$$p_1 = \frac{ah^2}{V_0}$$

Applications of McLeod Gauge

McLeod gauge is used mainly for calibrating other inferential type of gauges. The shortcomings of the McLeod gauge are its fragility and the inability to measure continuously. The vapor pressure of Mercury sets the lower limit of measurement range of the gauge.

Thermal Conductivity Gauges

The working principle of thermal conductivity gauges is that at low pressures heat lost by a heated object by conduction through molecules will depend on pressure. This is valid only for certain pressure range.

When the mean free path is comparable with the dimensions of the gauge head the heat loss from a heated wire in the gauge head will be by (i) conduction through leads (ii) radiation to surroundings (iii) conduction through molecules.

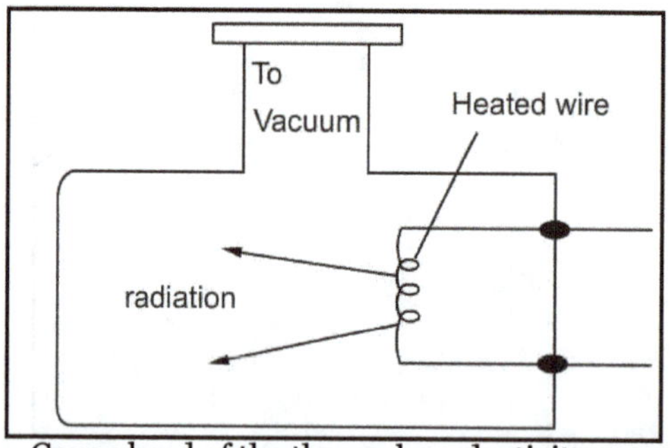

Gauge head of the thermal conductivity gauge.

The range of thermal conductivity gauges is from mm of Hg to 10^{-3}mmHg. At higher pressures, heat loss from the heated wire is insensitive to the pressure change. At lower pressure heat loss by (i) and (ii) become more significant. There are two kinds of the thermocouple gauges.

Pirani Gauge

Measures change in resistance of the heated wire when it looses heat to the gas molecules in the gauge head. In this case the gauge is called pirani gauge.

Pirani Gauge

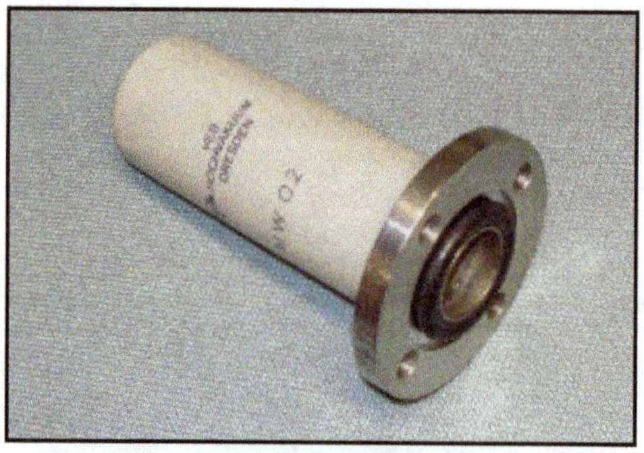
Pirani probe with 64 mm vacuum flange.

The Pirani gauge is a robust thermal conductivity gauge used for the measurement of the pressures in vacuum systems. It was invented in 1906 by Marcello Pirani.

Structure

The Pirani gauge consists of a metal filament (usually platinum) suspended in a tube which is connected to the system whose vacuum is to be measured. Connection is usually made either by a ground glass joint or a flanged metal connector, sealed with an o-ring. The filament is connected to an electrical circuit from which, after calibration, a pressure reading may be taken.

Mode of Operation

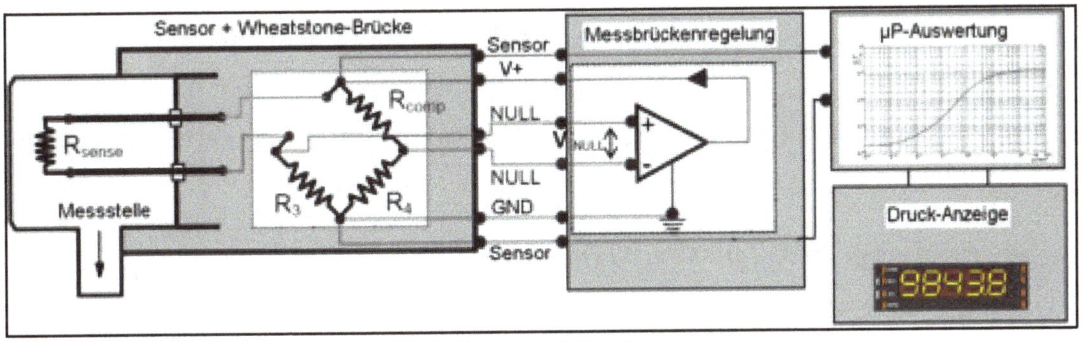

Block diagram of Pirani gauge.

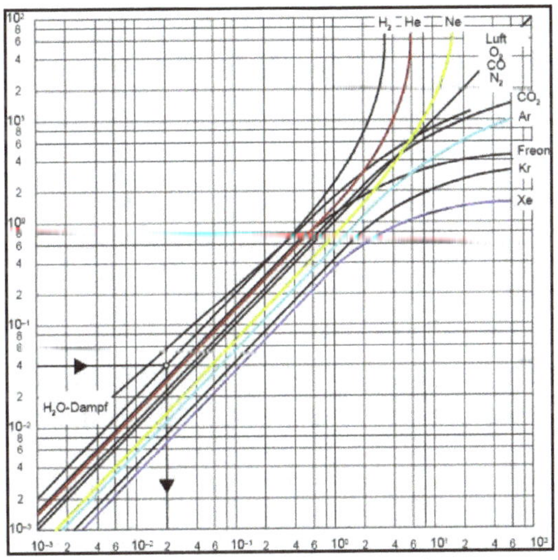

Curves to convert air readings to other gases.

A heated metal wire (also called a filament) suspended in a gas will lose heat to the gas as its molecules collide with the wire and remove heat. If the gas pressure is reduced the number of molecules present will fall proportionately and the wire will lose heat more slowly. Measuring the heat loss is an indirect indication of pressure.

The electrical resistance of a wire varies with its temperature, so the resistance indicates the temperature of wire. In many systems, the wire is maintained at a constant resistance R by controlling the current I through the wire. The resistance can be set using a bridge circuit. The power delivered to the wire is I^2R, and the same power is transferred to the gas. The current required to achieve this balance is therefore a measure of the vacuum.

The gauge may be used for pressures between 0.5 Torr to 10^{-4} Torr. The thermal conductivity and heat capacity of the gas may affect the readout from the meter, and therefore the apparatus may need calibrating before accurate readings are obtainable. For lower pressure measurement, the thermal conductivity of the gas becomes increasingly smaller and more difficult to measure accurately, and other instruments such as a Penning gauge are used instead.

Pulsed Pirani Gauge

A special form of the Pirani gauge is the pulsed Pirani vacuum gauge where the filament is not operated at a constant temperature, but is cyclically heated up to a certain temperature threshold by an increasing voltage ramp. When the threshold is reached, the heating voltage is switched off and the filament cools down again. The required heat-up time is used as a measure of pressure.

For adequately low pressure the following relation for supplied heating power and filament temperature $T(t)$ applies:

$$P_{\mathrm{el}} = C_1\lambda_{\mathrm{gas}}(T(t)-T_a)+C_2\lambda_{\mathrm{fil}}(T(t)-T_a)+A_{\mathrm{fil}}\epsilon\sigma(T(t)^4-T_a^4)+c_{\mathrm{fil}}m_{\mathrm{fil}}\frac{\mathrm{d}T}{\mathrm{d}t},$$

where c_{fil} is the heating capacity of the filament, m_{fil} is the mass of the filament and C_1 and C_2 are constants.

Advantages and Disadvantages of the Pulsed Gauge

Advantages

- Significantly better resolution in the range above 75 Torr.
- The power consumption is drastically reduced compared to continuously operated Pirani gauges.
- The gauge's thermal influence on the real measurement is lowered considerably due to the low temperature threshold of 80°C and the ramp heating in pulsed mode.
- The pulsed mode allows for efficient implementation of modern microprocessor technology.

Disadvantages

- Increased calibration effort.
- Longer heat-up phase.

Alternative

An alternative to the Pirani gauge is the thermocouple gauge, which works on the same principle of detecting thermal conductivity of the gas by a change in temperature. In the thermocouple gauge, the temperature is sensed by a thermocouple rather than by the change in resistance of the heated wire.

Thermocouple Gauge

Instead of measuring electrical resistance, a thermocouple is kept in contact with the heated wire and the temperature of the wire is directly measured as a measure of pressure.

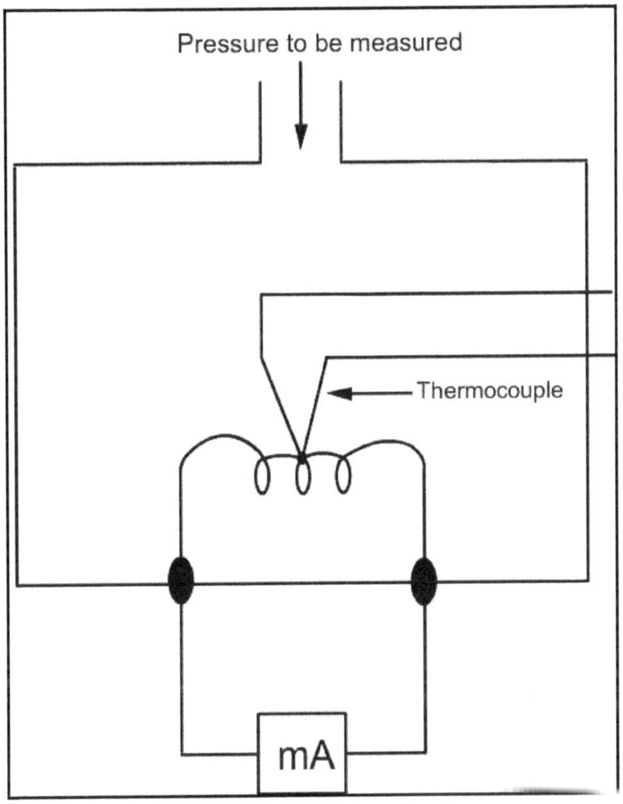

Thermocouple gauge head.

Hot-filament Ionization Gauge

The hot-filament ionization gauge, sometimes called a hot-filament gauge or hot-cathode gauge, is the most widely used low-pressure (vacuum) measuring device for the region from 10^{-3} to 10^{-10} Torr. It is a triode, with the filament being the cathode.

Function

A regulated electron current (typically 10 mA) is emitted from a heated filament. The electrons are attracted to the helical grid by a DC potential of about +150 V. Most of the electrons pass through the grid and collide with gas molecules in the enclosed volume, causing a fraction of them to be ionized. The gas ions formed by the electron collisions are attracted to the central ion collector wire by the negative voltage on the collector (typically −30 V). Ion currents are on the order of 1 mA/Pa. This current is amplified and displayed by a high-gain differential amplifier/electrometer.

This ion current differs for different gases at the same pressure; that is, a hot-filament ionization gauge is composition-dependent. Over a wide range of molecular density, however, the ion current from a gas of constant composition is directly proportional to the molecular density of the gas in the gauge.

Construction

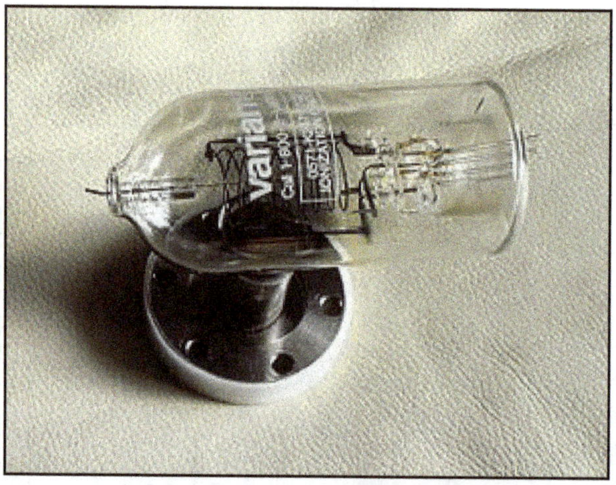

Tubulated hot-cathode ionization gauge.

A hot-cathode ionization gauge is composed mainly of three electrodes, all acting as a triode, wherein the cathode is the filament. The three electrodes are a collector or plate, a filament, and a grid. The collector current is measured in picoamperes by an electrometer. The filament voltage to ground is usually at a potential of 30 volts, while the grid voltage at 180–210 volts DC, unless there is an optional electron bombardment feature, by heating the grid, which may have a high potential of approximately 565 volts. The most common ion gauge is the hot-cathode Bayard–Alpert gauge, with a small collector inside the grid. A glass envelope with an opening to the vacuum can surround the electrodes, but usually the nude gauge is inserted in the vacuum chamber directly, the pins being fed through a ceramic plate in the wall of the chamber. Hot-cathode gauges can be damaged or lose their calibration if they are exposed to atmospheric pressure or even low vacuum while hot.

Electrons emitted from the filament move several times in back-and-forth movements around the grid before finally entering the grid. During these movements, some electrons collide with a gas molecule to form a pair of an ion and an electron (electron ionization). The number of these ions is proportional to the gas molecule density multiplied by the electron current emitted from the filament, and these ions pour into the collector to form an ion current. Since the gas molecule density is proportional to the pressure, the pressure is estimated by measuring the ion current.

The low-pressure sensitivity of hot-cathode gauges is limited by the photoelectric effect. Electrons hitting the grid produce X-rays that produce photoelectric noise in the ion collector. This limits the range of older hot-cathode gauges to 10^{-8} Torr and the Bayard–Alpert gauges to about 10^{-10} Torr. Additional wires at cathode potential in the line of sight between the ion collector and the grid prevent this effect. In the extraction type the ions are not attracted by a wire but by an open cone. As the ions cannot decide which part of the cone to hit, they pass through the hole and form an ion beam. This ion beam can be passed on to a:

- Faraday cup,
- quadrupole mass analyzer with Faraday cup,
- microchannel plate detector with Faraday cup,
- quadrupole mass analyzer with microchannel plate detector Faraday cup,
- ion lens and acceleration voltage and directed at a target to form a sputter gun; in this case a valve lets gas into the grid cage.

Types of Hot-filament Ionization Gauges

- Bayard–Alpert (uses sealed tube).
- Nude gauge (uses the vacuum chamber to make a complete seal).

Hot Cathode Ionisation Gauges

At higher levels of vacuum, the measurement of pressure is by using Ionisation gauges. They operate on the principle of ionising the gas by means of electrons emitted by a heated filament. The kinetic energy acquired by an electron in passing through a potential difference of V volts corresponds to a value equal to V*e where e is the charge of the electron. When this energy exceeds a certain critical value corresponding to the ionization potential Vi, there is a possibility that collisions between molecules and electrons will result in the formation of +ve ions. The relatively high velocity electrons on hitting a gas molecule drives an electron out of it leaving it positively charged. For gases such as N_2, O_2 etc, Vi is ~ 15volts. The measurement of the ions produced is a measure of the pressure. The electrons are speeded up by an electric field and +ve ions produced are collected. The number of +ve ions formed will depend on the number of molecules and therefore on the pressure.

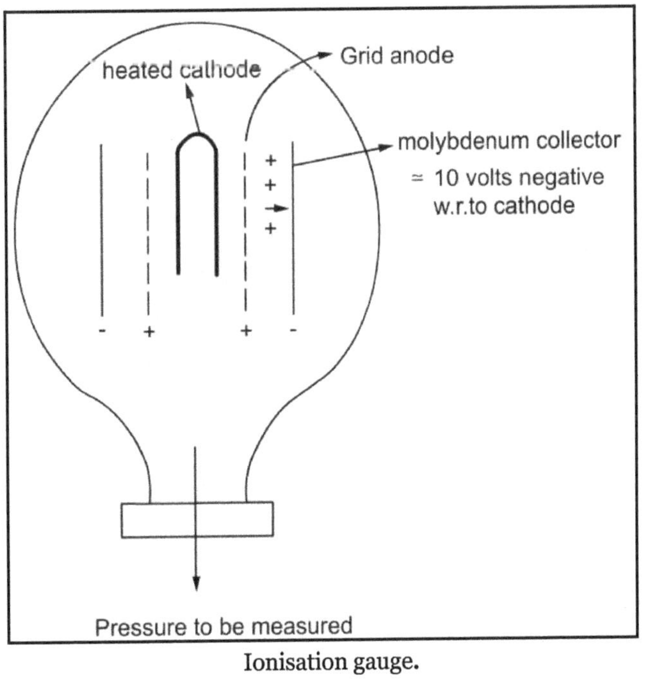

Ionisation gauge.

The gauge consists of a cathode, grid anode and a negative plate. The negative plate is at ~10V negative with respect to the cathode. The electrons emitted by the hot cathode (filament) are speeded up by the electric field and the positive ions produced are collected by the negative plate.

Cold Cathode Ionisation Gauge (Penning Gauge)

This gauge also works on the ionization principle. Positive ions are produced by the electrons and current due to these ions gives a measure of the pressure. Electrons are ejected from a cold cathode of Zirconium, Thorium by electric discharge. The gauge consists of two plate cathodes and a ring anode . A potential difference of ~2KV is applied across thc electrodes.

The travel of electrons is made over a much longer distance. The secondary electrons are made to travel in helical paths before reaching the anode. This is accomplished by a magnetic field.

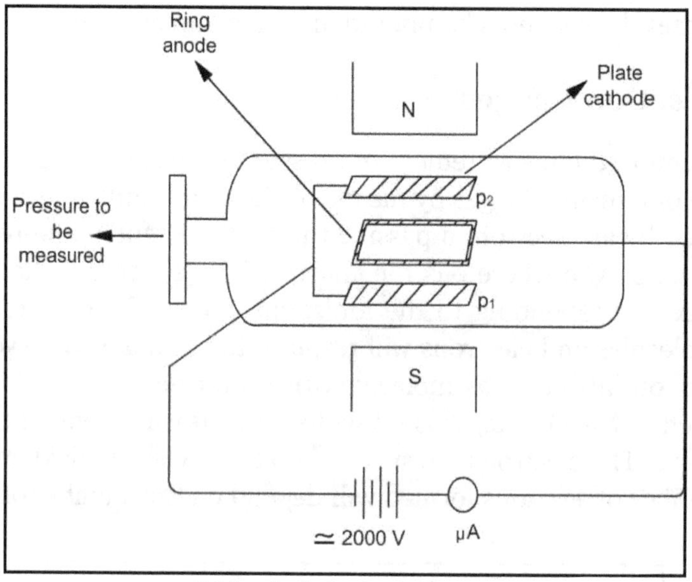

Cold cathode ionization gauge.

Magnetic poles are kept such that the flux with lines of force is applied perpendicular to the two cathodes.

Measurement of Pressure in Flows

In flows, distinction has to be made between static and stagnation pressures.

Static Pressure

Pressure acting on the surface of a body imagined to be moving with the fluid with the same velocity as the medium is the static pressure.

Stagnation Pressure

It is the pressure of a fluid imagined to be brought to rest isentropically.

Measurement of Static Pressure

Common technique is to connect a probe to an orifice drilled perpendicular to the wall of the model where the streamlines are undistorted and parallel.

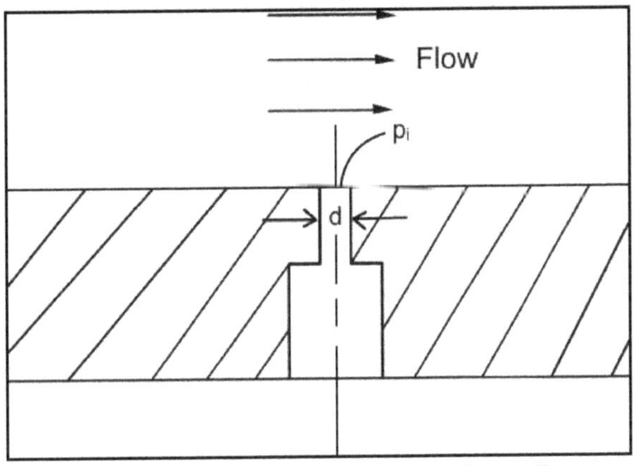

Static pressure orifice (tap) on a flat wall.

Static hole diameter is about 1/5 of the boundary layer thickness. Practically the diameter is about 0.25mm on small models and 2.5mm on larger installations. The correctness of the static pressure measured is dependent on 'd' the orifice diameter. The influence of the orifice diameter on the error in static pressure measurements is given graphically in the figure given below.

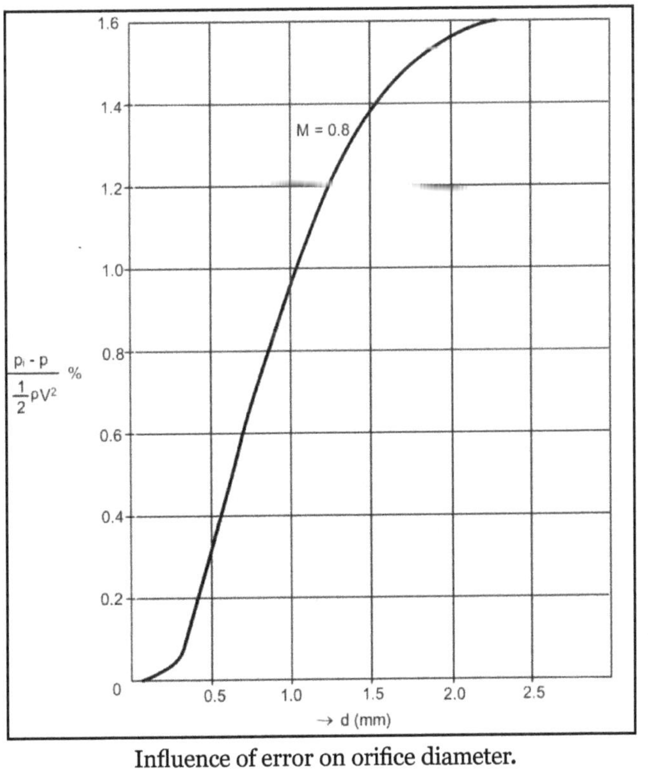

Influence of error on orifice diameter.

The static pressure orifices (taps) on the walls of the flow channel provide the wall static pressure. The wall static pressure can not be assumed to prevail inside the flow. In order to get the pressure inside the flow, probes of suitable design have to be made use of.

Static Pressure Probes for Subsonic Flow

The commonly used static pressure probe in subsonic flows is the Prandtl probe. The Prandtl probe is an intrusive device. The pressure sensing orifices on the periphery of the probe are carefully located such that the influence of the nose and stem of the probe nullify each other.

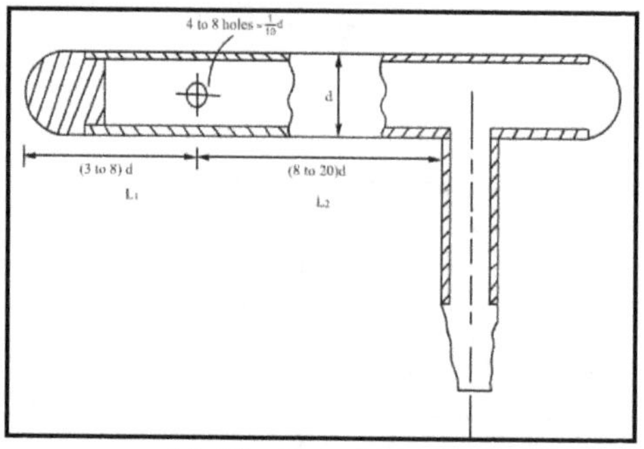

(a) The Prandtl probe.

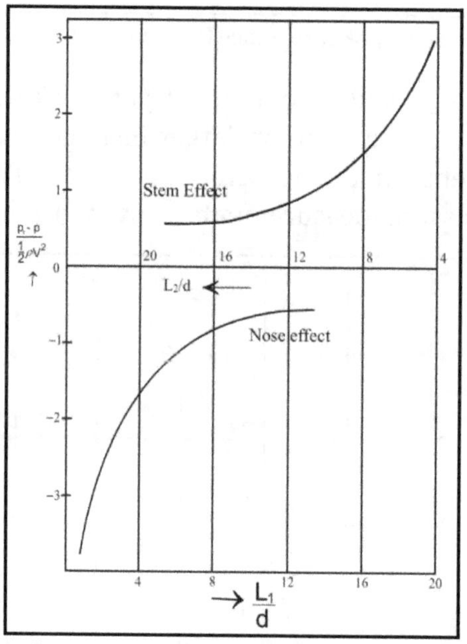

(b) The nose and stem effects in a Prandtl tube.

Because of the intrusive nature of the probe, the flow will be accelerated by the nose. The effect will be to reduce the static pressure which is called the nose effect. In contrast, the effect of stem will be to locally stagnate the flow and thereby to increase the static pressure. This effect is the stem effect. As shown in the fig. (b), the position of the orifices are so chosen that the two effects neutralize each other. Additionally the probe must be slender (say ~ 1.0 to 1.5mm dia) and kept parallel to the flow.

Static Pressure Probes for Supersonic Flow

When Mach number is more than 1.0, shock appears. When the cone angle of the probe is less than the shock detachment angle for the given Mach number shown in fig. (c) and if the orifices are located well downstream of the shock wave, the measured static pressure will tend towards the value for the undisturbed flow. Conical or ogival shaped tubes are used.

Probes are made small and with static taps on the cone surface. The effect of yaw is reduced by arranging several orifices so that the pressure inside the tube is an average value. Usually the tube has 4 to 8 orifices whose diameter is about 1/10 th of the outside diameter of the tube.

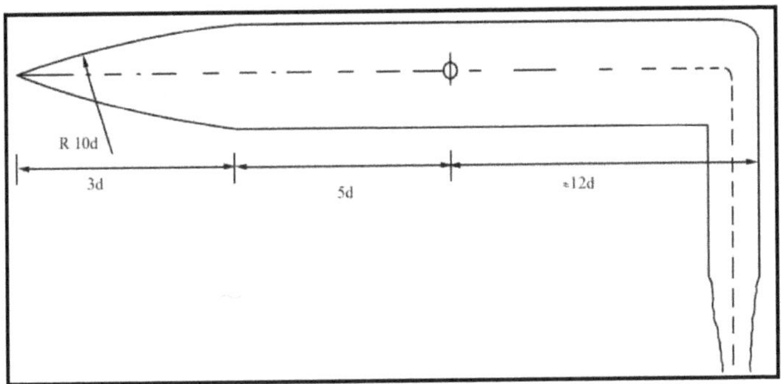

(a) Ogival shaped static pressure probe for supersonic flow.

Good results are obtained with ogival tubes. The tube shown in figure above has a systematic error within 1%. Angle of the cone should be less than the angle at which the shock wave becomes detached from the cone. The error in the measurement of static pressure decreases as the distance of the orifice from the tip of the probe increases.

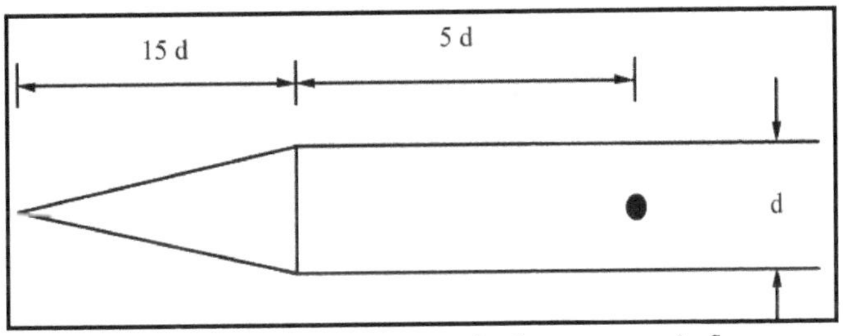

(b) Conical static pressure probe for supersonic flow.

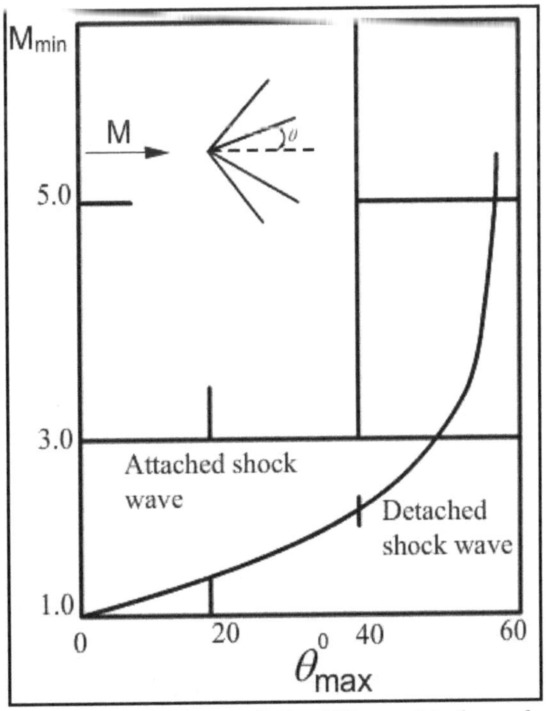

(c) Maximum deflection angle for different Mach numbers.

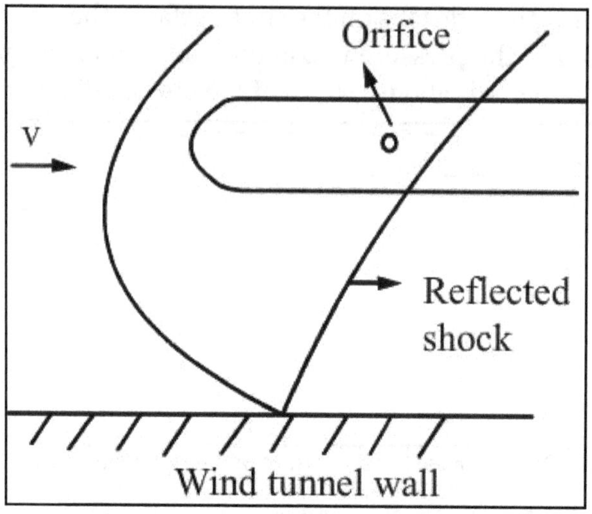

(d) Reflection of the shock from the wall.

It is important to have long pointed tubes. Otherwise there is the possibility that the reflected shock may affect the static pressure readings. Roughness at the edges of the orifices too may cause large errors in measurement.

Measurement of Stagnation Or Total Pressure

Insubsonic Flows

The gas particles come to rest so quickly at the stagnation point of a body, that heat transfer and friction losses are negligible. Therefore, in subsonic flows only isentropic changes occur. Hence the stagnation pressure measured in a subsonic flow is not significantly different from that in the settling chamber. Total pressure is measured with a Pitot tube which a cylindrical tube is having an orifice pointing towards the flow. Axis of tube must coincide with the flow direction. It is a practice to use tubes which generally have blunt ends.

Such tubes are insensitive to yaw upto ± 10 to 12°

The relation between the stagnation pressure measured and the static pressure may be expressed as follows:

$$\frac{p}{p_0} = \left(1 + \frac{r-1}{2} M^2\right)^{\frac{r}{r-1}}$$

where p_0 is the total pressure and p the static pressure.

Insupersonic Velocities

Shockwave appears upstream of the tube nose. Therefore tube measures only pressure behind the shock wave. Normal shock equations give the relation between the total pressures upstream and downstream of the shock. As the bow shock formed in front of the Pitot tube has the normal part only in the central region, the tube diameter for measurements in supersonic flow is usually kept very small.

Lag in Manometric Systems

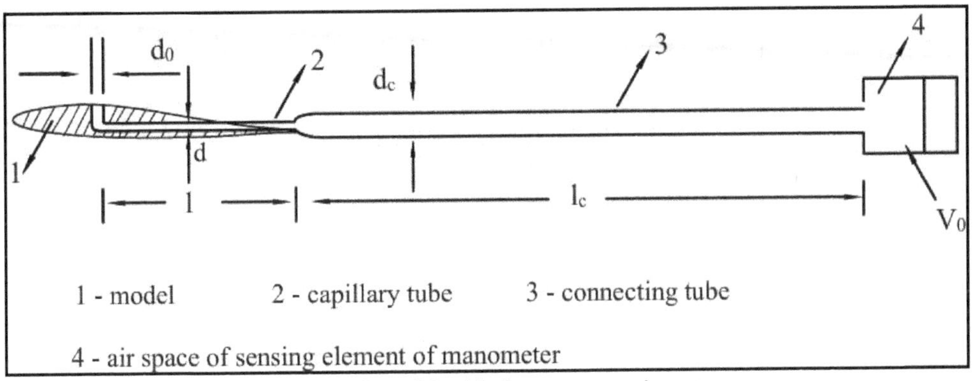

Wind tunnel model with the manometric system.

When the pressure changes near the orifice which is connected to a manometer, equilibrium in the manometer is established not immediately but after a certain time. If the pressure is read of earlier there will be gross errors.

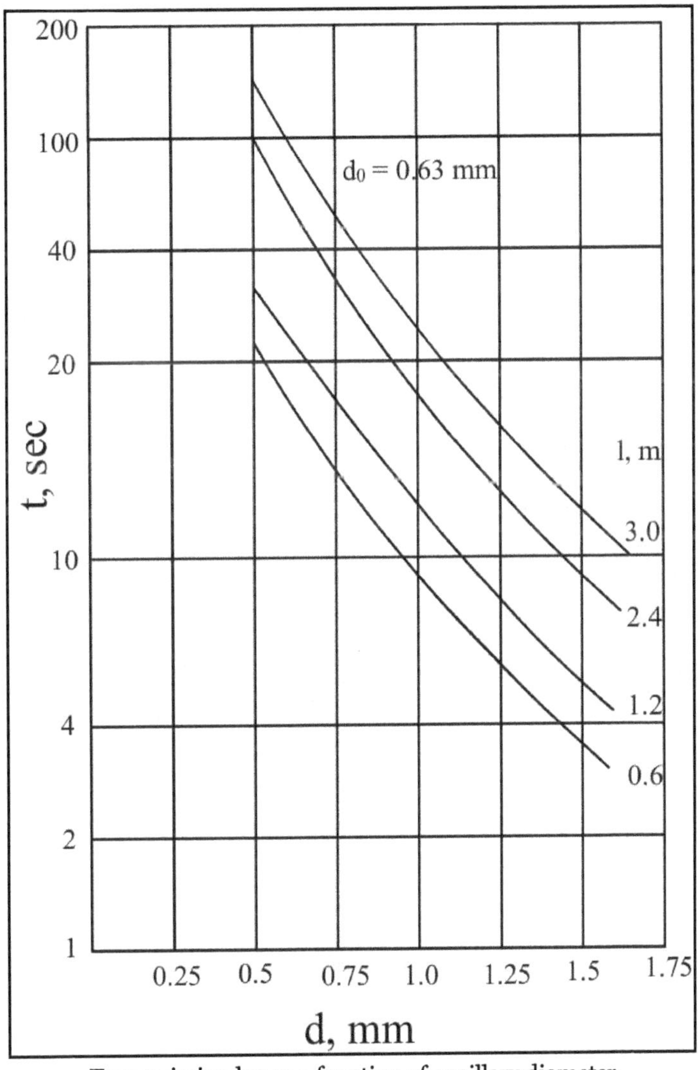

Transmission lag as a function of capillary diameter.

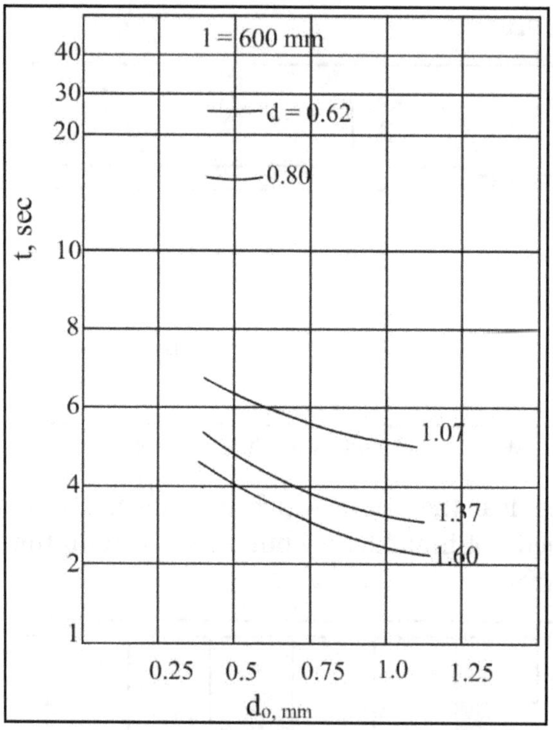

Transmission lag as a function of orifice diameter.

Smaller transmission lags are necessary not only for high reliability but also for reducing the duration of experiments. Equilibrium will be established in the manometric system later after the pressure on the model is stabilized. The run times should be longer than the transmission lag.

Transmission lag is caused by:

a) Resistance of the tubes

b) The change in air density

c) and inertia of the moving masses

The main factors influencing the transmission lag are the orifice diameter d_o, the diameter d of the capillary and d_c of the connecting tube and their respective lengths l and l_c .

The orifice diameter is of small influence when $d/d_o < 2.5$. When $d/d_o > 2.5$ the transmission lag increases sharply. The orifice diameter should not be less than half the diameter of the capillary tube. The influence of the diameter of the capillary tube is very strong. A reduction of this diameter has its main effect an increase in the resistance to the flow of gas. An increase in the length of the capillary tube has a significant effect on the lag. Capillary tubes should have larger diameter and shorter length.

Permissions

All chapters in this book are published with permission under the Creative Commons Attribution Share Alike License or equivalent. Every chapter published in this book has been scrutinized by our experts. Their significance has been extensively debated. The topics covered herein carry significant information for a comprehensive understanding. They may even be implemented as practical applications or may be referred to as a beginning point for further studies.

We would like to thank the editorial team for lending their expertise to make the book truly unique. They have played a crucial role in the development of this book. Without their invaluable contributions this book wouldn't have been possible. They have made vital efforts to compile up to date information on the varied aspects of this subject to make this book a valuable addition to the collection of many professionals and students.

This book was conceptualized with the vision of imparting up-to-date and integrated information in this field. To ensure the same, a matchless editorial board was set up. Every individual on the board went through rigorous rounds of assessment to prove their worth. After which they invested a large part of their time researching and compiling the most relevant data for our readers.

The editorial board has been involved in producing this book since its inception. They have spent rigorous hours researching and exploring the diverse topics which have resulted in the successful publishing of this book. They have passed on their knowledge of decades through this book. To expedite this challenging task, the publisher supported the team at every step. A small team of assistant editors was also appointed to further simplify the editing procedure and attain best results for the readers.

Apart from the editorial board, the designing team has also invested a significant amount of their time in understanding the subject and creating the most relevant covers. They scrutinized every image to scout for the most suitable representation of the subject and create an appropriate cover for the book.

The publishing team has been an ardent support to the editorial, designing and production team. Their endless efforts to recruit the best for this project, has resulted in the accomplishment of this book. They are a veteran in the field of academics and their pool of knowledge is as vast as their experience in printing. Their expertise and guidance has proved useful at every step. Their uncompromising quality standards have made this book an exceptional effort. Their encouragement from time to time has been an inspiration for everyone.

The publisher and the editorial board hope that this book will prove to be a valuable piece of knowledge for students, practitioners and scholars across the globe.

Index

www.ingramcontent.com/pod-product-compliance
Lightning Source LLC
LaVergne TN
LVHW080717210926
848485LV00003B/483